VVIP 高端会所设计

VVIP Top Club Design

DAM 工作室　主编
Michelle　翻译

華中科技大學出版社
http://www.hustp.com
中国 · 武汉

图书在版编目（CIP）数据

VVIP 高端会所设计 / DAM 工作室 主编 . – 武汉：华中科技大学出版社，2013.5

ISBN 978-7-5609-8958-7

Ⅰ . ① V… Ⅱ . ① D… Ⅲ . ①服务建筑—建筑设计 Ⅳ . ① TU247

中国版本图书馆 CIP 数据核字（2013）第 102694 号

VVIP 高端会所设计　　DAM 工作室 主编

出版发行：华中科技大学出版社（中国・武汉）
地　　址：武汉市武昌珞喻路 1037 号（邮编：430074）
出 版 人：阮海洪

责任编辑：熊纯　　责任监印：秦英
责任校对：王莎莎　　装帧设计：筑美空间

印　　刷：中华商务联合印刷（广东）有限公司
开　　本：965 mm × 1270 mm 1/16
印　　张：19
字　　数：152 千字
版　　次：2013 年 8 月第 1 版 第 1 次印刷
定　　价：318.00 元（USD 59.99）

投稿热线：（020）36218949　　1275336759@qq.com
本书若有印装质量问题，请向出版社营销中心调换
全国免费服务热线：400-6679-118 竭诚为您服务

荣耀的空间

—— 会所设计的告白

文 / 刘卫军

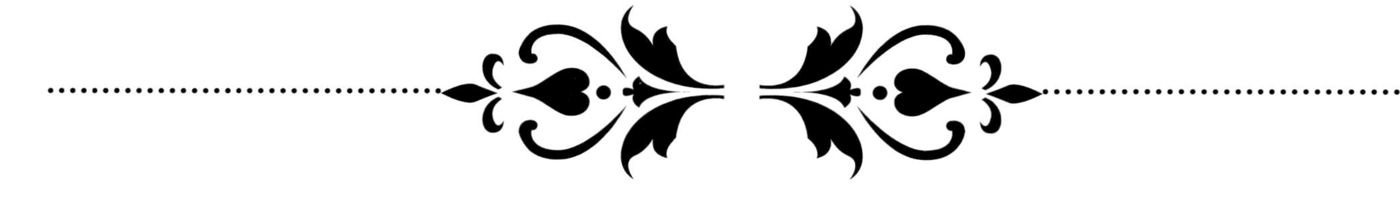

荣耀的空间——会所，我之所以这样认为是因为会所表达的不仅仅是平面或立体的空间设计，还是商业与艺术相融合的空间设计。

我曾经将“灵感触发”作为会所设计的思想，表达那份期待与回忆，反映我对会所空间设计的一种态度与情愫，充满激情地将多元化的商业精神与艺术融会贯通到作品的每个空间、每个环节。一种微妙的思想总在触动着空间中感性与理性的变化，彰显主题性的创意思维，以独特的气质凸显商业艺术与空间交融的撼人魅力。

近年，我国商品经济迅速发展，这不但影响了我们的审美观，也冲击了消费者追求情感的思维平衡点。这一切的变化随着社会的进步与城市的发展，形成了一种精神消费的文化现象，也促进了市场经济与服务行业的发展与转化。由此，“会所”一词应运而生。

会所，属服务行业，与酒店业同类。随着时间的发展，已由单一附属经营发展为综合多元化独立经营。由于会所经营具有高度的私享性，因此倍受少数知性名流、成功人士的追捧，也成为了聚会、交流、分享的平台。马会、游艇会、名流会、高尔夫会所、高端楼盘会所、SPA 养生保健会所、私人派对会所、娱乐运动会所、企业家商务会所、特色餐饮会所等各类主题性及功能性会所纷纷涌现。尽管主题性及功能性会所空间的设计风格与功能诉求各不相同，但它的“荣耀”特性却是会所空间永恒的追求。因为会所空间不再是点面的简单组合，而是一种“圈子型”的商业文化现象的表现。设计中我们不仅不能忽视它的感染调性、形式格调与舒适体验性，同时还要将文化艺术融于空间，创造无限商业艺术感染力，提升环境的体验价值，满足客人的尊荣感与私享性，让心理与生理备受尊宠。

此次相聚这里，希望通过本书传递一点我个人的实践经验，以及关于会所设计的心声，让我们共同分享、共同创造更多传奇、完美的“荣耀空间”。

目录 CONTENTS

Fusion Club

西水融会会所

Design Company: Wuxi S-zona Design and Production Co., Ltd.
Designer: Feng Jiayun
Participants: Tie Zhu, Geng Shunfeng
Area: 1,440 m^2
Main Materials: Old Wood, Machruian Ash, Black Stone, Copper Wire Drawing, Horse Fur, Sound-absorbing Board

设计公司：无锡市上瑞元筑设计制作有限公司
项目设计：冯嘉云
参与设计：铁柱、耿顺峰
项目面积：1 440 m^2
主要材料：老木板、水曲柳染色、黑洞石、拉丝铜、马毛皮、木丝吸音板

A bright, thick historical memory and gleamed, classical, shady space, correspond with the temperament of "China modern industry and commerce" and "Republican period" with glorious history. The original intention of the project is to shape the narrative of decoration. At the same time, designers also pay attention to make intellectual and stylish space, to establish a high level taste in corresponding to the target customers' expectations.

The club is destined to be a harbor for body and mind of a small group rich people in the city. Therefore, the designers used international greys to represent noble spirit in natural, stable and generous space. The black leather, grey blue wallpaper, cloth, grey watered stone, magnificent natural grain of wood, camel carpet, brown chair and dark yellow cowhide are deducing the natural transition from cold tone to be warm tone. The contrast of ample materials, and the change of lively lines formed the tension of space.

斑驳、古意、婆娑的空间肌理，带有鲜明、厚重的历史印记，与曾经辉煌的“中国近现代工商业”、“民国”等语汇在气质上吻合。塑造故事性，成为本案设计的初衷。同时，知性、格调感的空间，亦应建立在满足高端目标群体的心理预期的基础上。

会所，注定是一小族群体的身心归所，是城市新贵“后奢侈、慢生活”的专属现场。为此，在色调上，采用国际化手法表现的灰调，于浑然天成、沉稳大气的空间中暗含对贵族精神的关照。从黑的皮革，灰蓝的墙纸、布艺，灰色水纹的石材，到瑰丽大方的木纹，驼色的地毯，褐色的椅背、桌套及深黄的牛皮，演绎着由冷色调到暖色调的自然过渡，丰富的材质对比和生动的纹饰变化形成了空间张力，内敛中流溢出悦动。

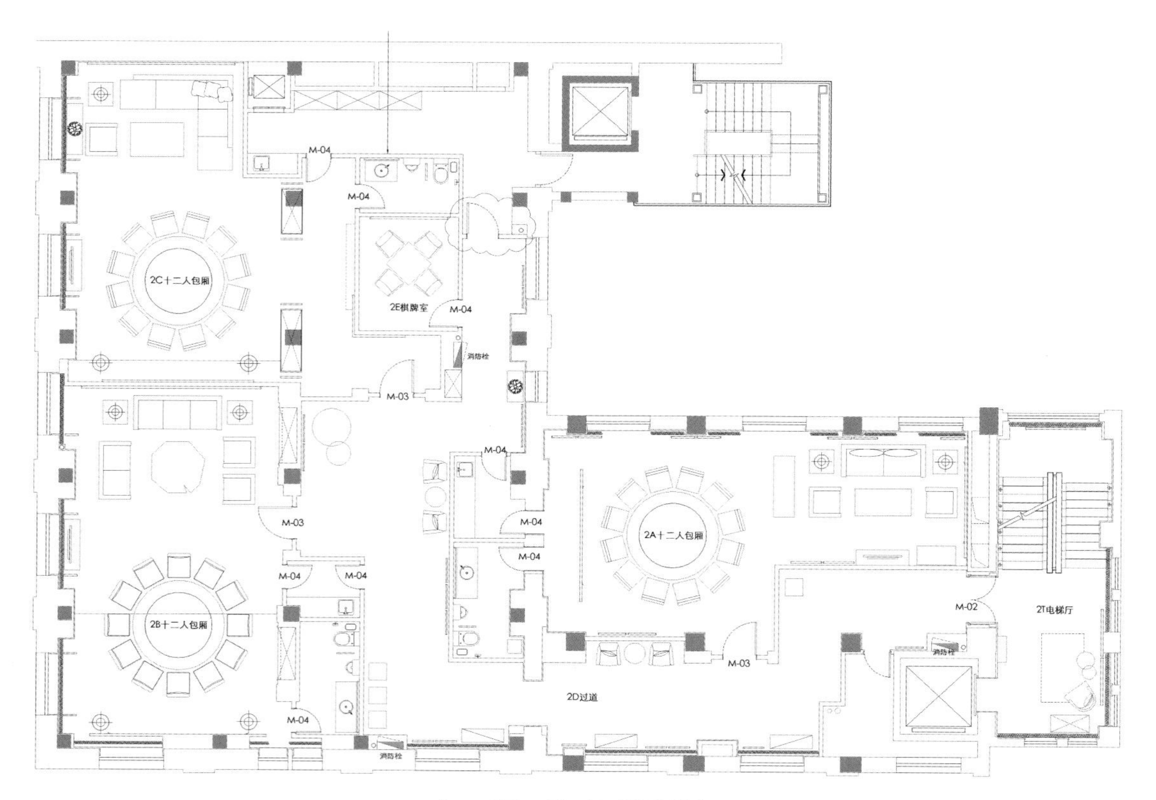

Second Floor Plan / 二层平面图

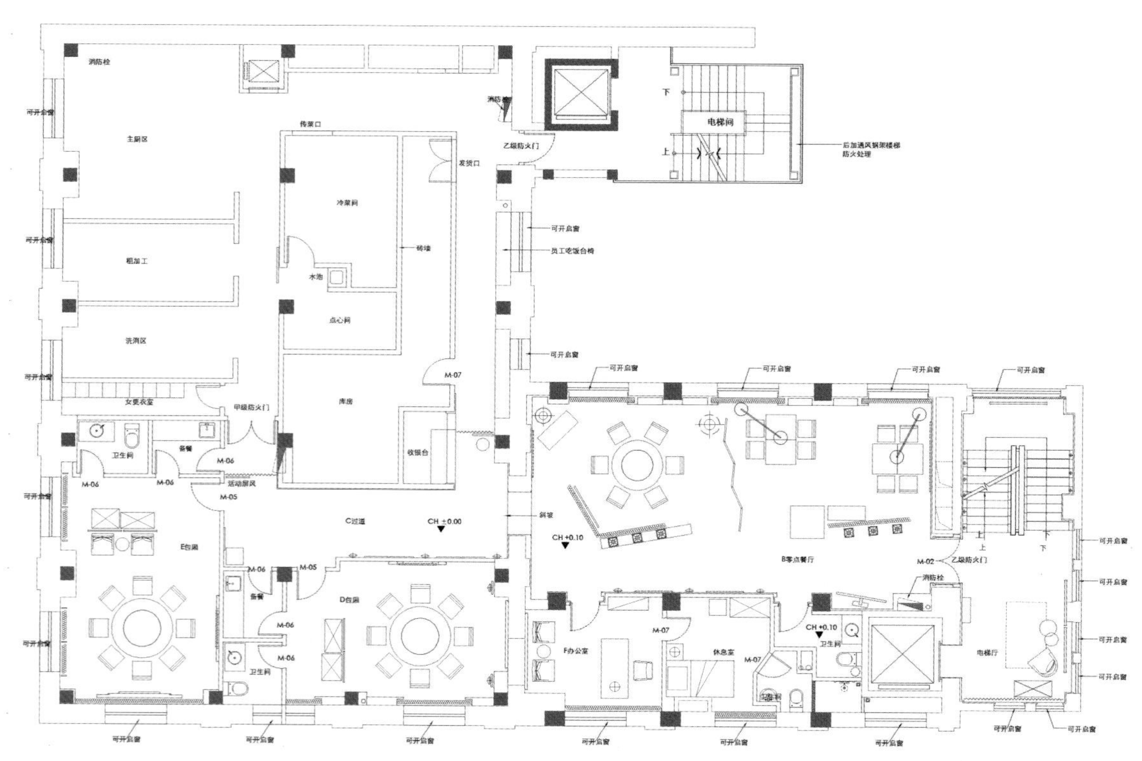

Third Floor Plan / 三层平面图

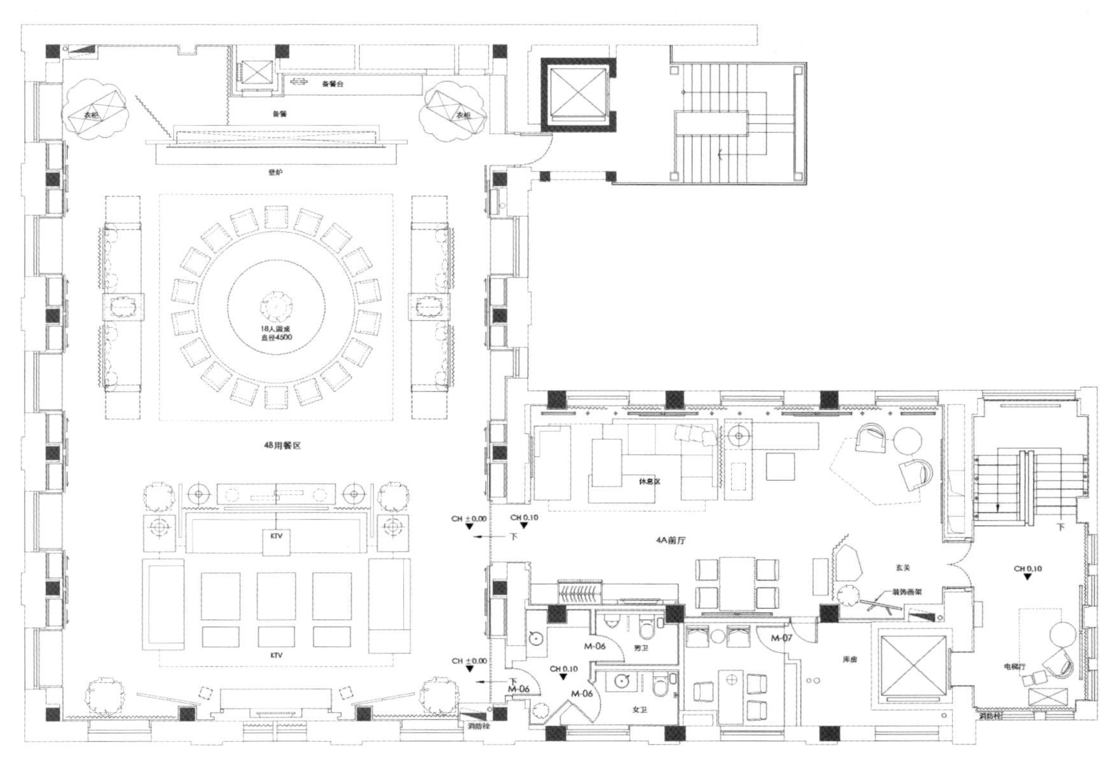

Fourth Floor Plan / 四层平面图

Yinxiang Club

扬州印象足道会所

Design Company: Wuxi S-zona Design and Production Co., Ltd.
Designer: Sun Liming
Participants: Geng Shunfeng, Chen Hao, Zhu Hengsong
Area: 1,000 m^2
Main Materials: Stone, Straw Wallpaper, Ashtree Opening Paint

设计公司：无锡市上瑞元筑设计制作有限公司
项目设计：孙黎明
参与设计：耿顺峰、 陈浩、朱恒松
项目面积：1 000 m^2
主要材料：石材、草编墙纸、水曲柳开口漆

The theme of the space is essence of water. It is symbolized by different marks, which brings elegant and comfortable feeling to people with brown, yellow and white natural gradient. The space was endowed with linear rhythm and harmonious connection of each functional space by designers using cutting method of spatial structure. The balance of space, the contrast of reality and fantasy, the match of light and heavy, the interaction of entity and light and shadow, build a zen significant implication of south-east Asia. On interpretation of elements, the water wave on the wall is performance of contemporary form, and clearly suggests the theme of "water". Reed screen, ink spun silk yarn, ethereal classical landscape curtain, not only added cultural temperament to the space, but also brought many different visual and psychological feelings to people, and created a clean, bright and elegant leisure space.

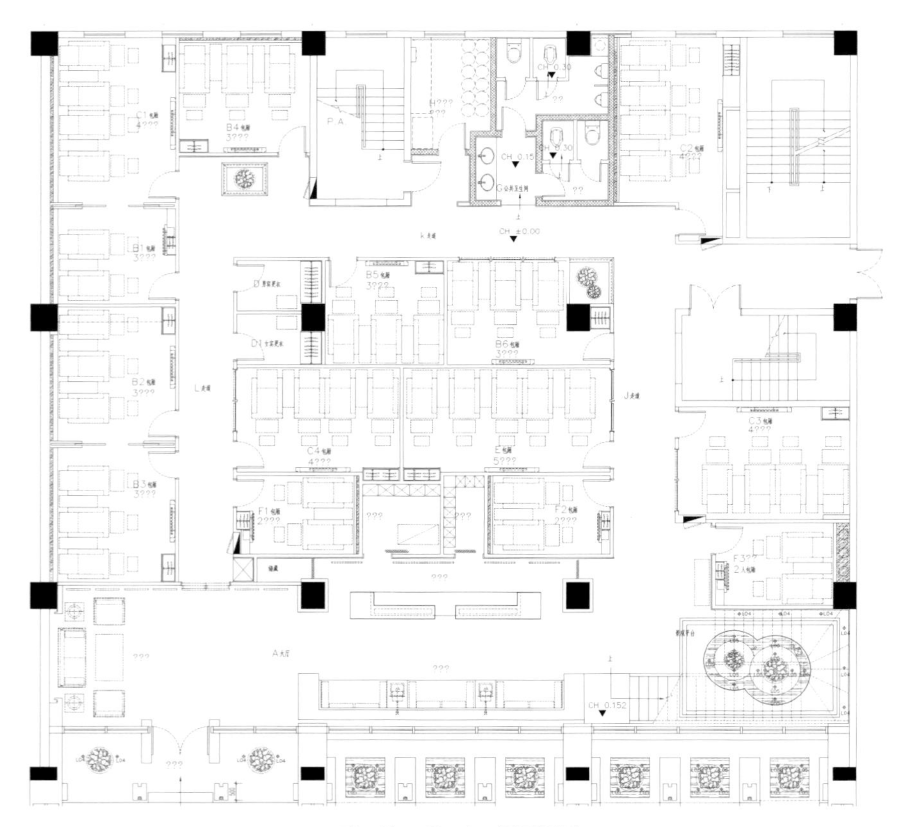

First Floor Plan / 一层平面图

以水的意象作为空间主题，与业态的属性保持形神的呼应，并对水的意向做了巧妙的符号化处理，自然化的褐、黄、白色渐进地渲染出空间的清雅高贵、舒适惬意。设计师利用空间结构的切割手法，赋予了空间线性的律动和各功能空间的和谐相连，空间上下部的均衡、虚实空间的对比、轻巧与朴拙的搭配、实体与光影的互动，营造出禅味深长的东南亚意蕴。在元素的演绎上，墙体极具浮雕感的水波既为形式上的当代化表现，又鲜明地暗示着“水”这一主题意象；苇帘、水墨的绢纱、飘渺的古典山水挂帘，不仅平添了空间的文化气质，又在材质与形式上丰富了视觉与心理感受，营造出一个净爽、淡雅的休闲空间。

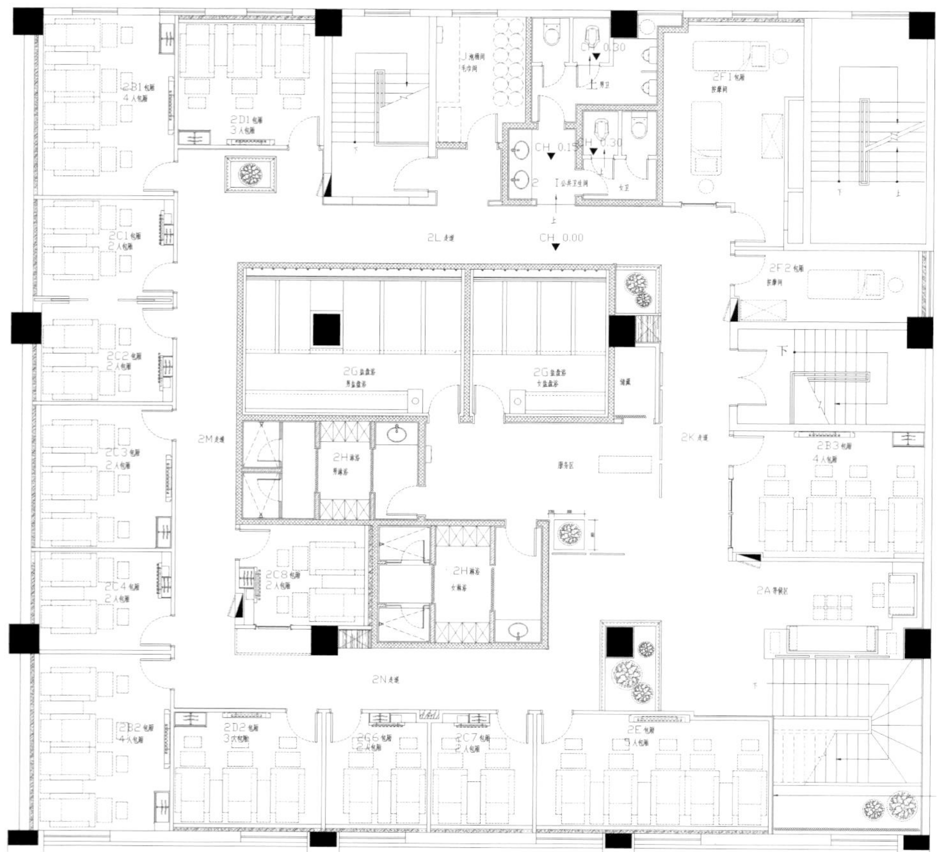

Second Floor Plan / 二层平面图

安全出口
EXIT

217

217

Quality·Light World
上品・光聚汇

Design Company: Daohe Design
Designer: Zhang Yundeng
Photographer: Li Lingyu
Area: 180 m^2
Main Materials: Emulsifying Glass, Red Oak Veneer, Dark Grey Plastic Board, Ariston Marble, Grey Mirror, White Painting Glass

设计公司：道和设计
项目设计：张云灯
项目摄影：李玲玉
项目面积：180 m^2
主要材料：乳化玻璃、红橡木饰面板、深灰色塑胶地板、雅士白大理石、灰镜、白色烤漆玻璃

It is a display space decorated by mirror which is the core element in the project. The reflected light and shadow provide a mysterious atmosphere by arranging mirrors reasonable. Through the corridor decorated with mirrors, looking at the space reflecting the surrounding scene in the mirror, it's hard to tell what is reality and imagination. Decorating by reflected mirrors makes the space more open and elegant. Coming to the main hall in the warm light, a functional area with simple and elegant layout is founded. It decorated with white frosted glass wall and dark grey plastic board. A long wooden table and Chinese style armchairs are placed in the central space. The simple natural wood color full in the space. The long table is set behind a Chinese style latticed window decorated with frosted glass and spotlights. When the spotlights turn on, the light will be through the frosted glass. This decoration is full of rich thinkings of the designer. Slightly is put on the other side of the space of a set of modern black leather sofa with special texture, and a tea table with black marble texture is placed behind it. A white chair brings vigour to this slightly depressing tonal space. Light and glass are the theme of this design. The combination of light and glass bring different flavors to the space, and ubiquitous sculptures with characteristics are sending out feeling of art.

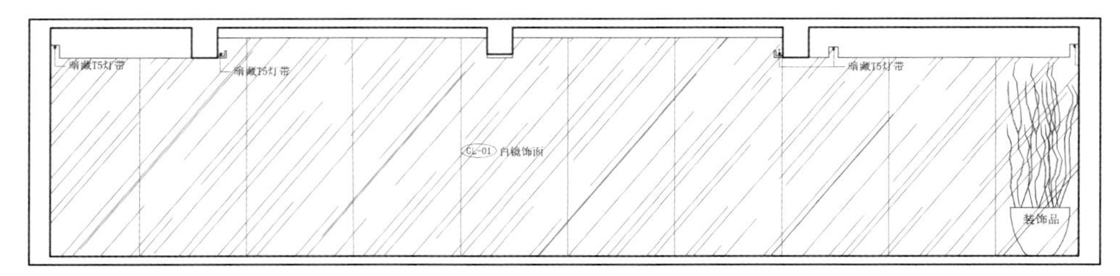

Elevation Drawing / 立面图

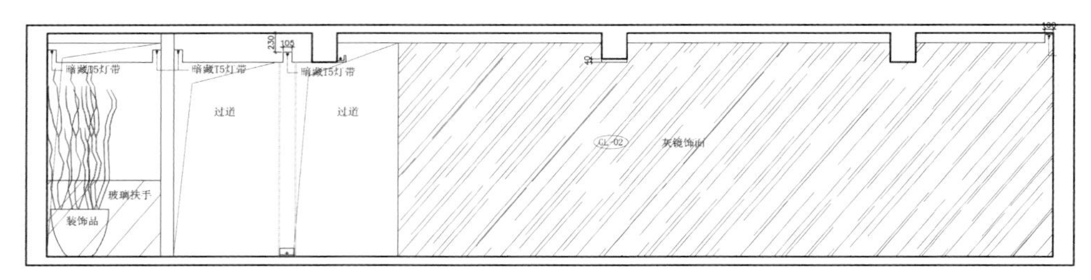

Elevation Drawing / 立面图

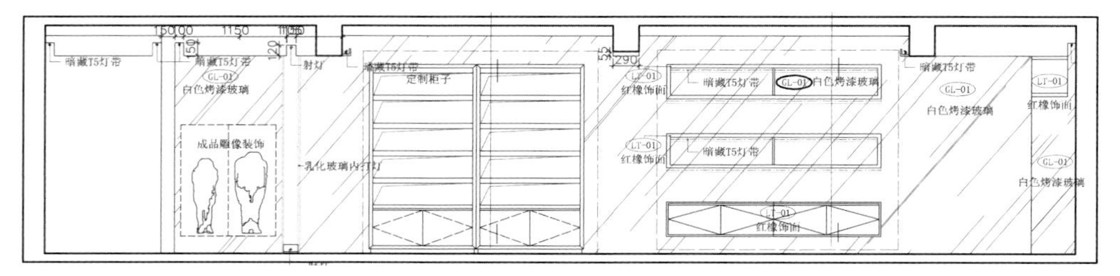

Elevation Drawing / 立面图

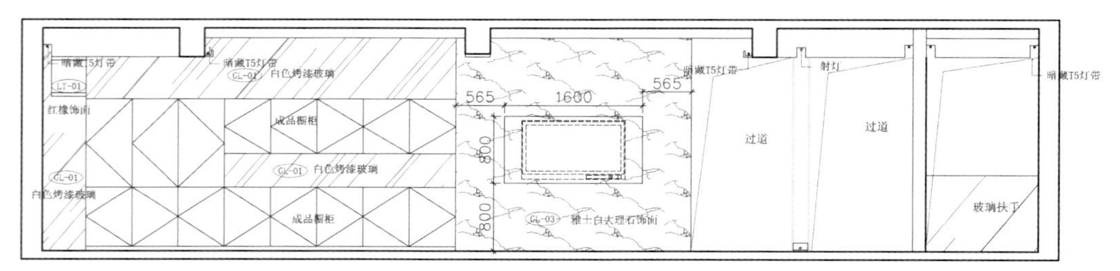

Elevation Drawing / 立面图

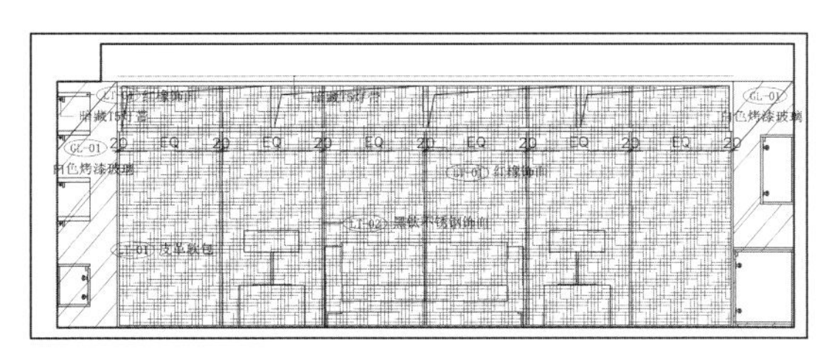

Elevation Drawing / 立面图

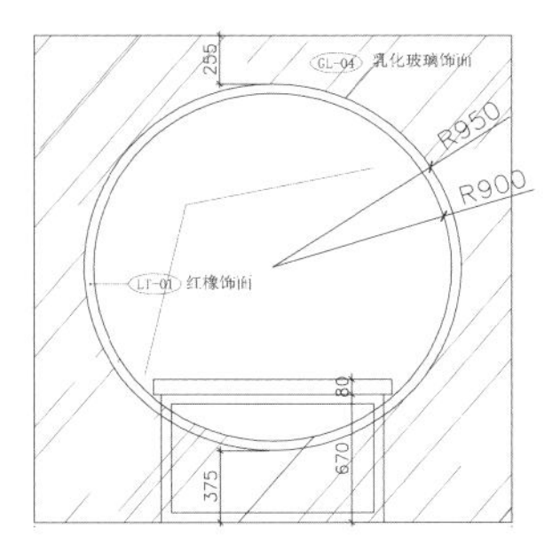

Elevation Drawing / 立面图

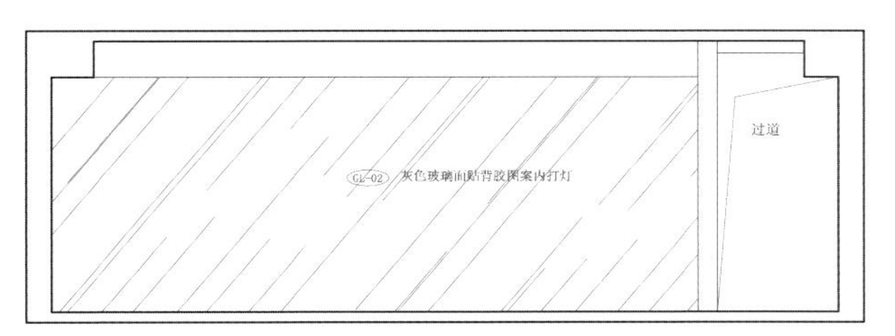

Elevation Drawing / 立面图

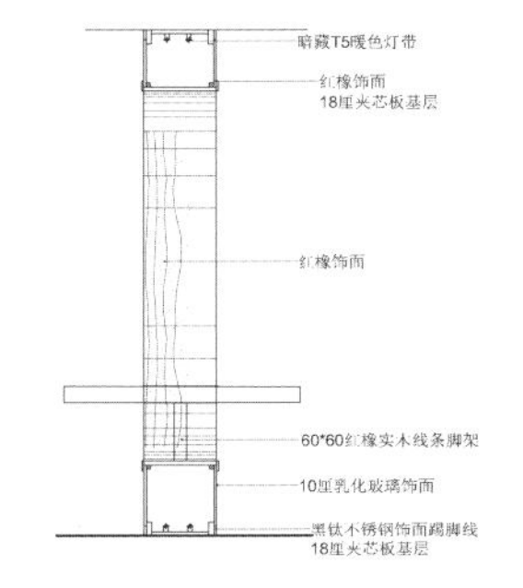

Sectional Drawing / 剖面图

这是一个以镜面为核心元素的展示空间，设计师将镜面进行合理的布置，并通过其反射与投影，使空间有着一种神秘之感，让人产生一种一探镜中世界的强烈愿望。置身用镜面布置的走廊里，镜中影像，已难于区分真实与虚幻。镜子的反射使空间少了些压抑，多了些大气。在射灯散发的柔光中，主厅显得十分素雅，白色的墙壁搭配上深灰色的塑胶地板，墙壁则用磨砂玻璃作为装饰。中部空间布置着木质长桌，配上中式圈椅，古朴自然的木色溢满空间。长桌穿过一个用磨砂玻璃做成的中式花窗，花窗内部埋设射灯，当灯被打开时，灯光便会透过磨砂玻璃散发出来，十分富有巧思。空间的另一侧则摆放着略显现代的黑色皮质沙发套件，其特色的纹路，舒适感十足。而黑色大理石纹理的茶几，则带着几分深沉，一旁的白色球椅给这略显沉闷的色调带来一丝活力，缓和了空间的气场。灯光与玻璃是这个空间的主题，它们于此互相交融，使空间极具风味，而随处可见的特色雕塑也散发着艺术气息。

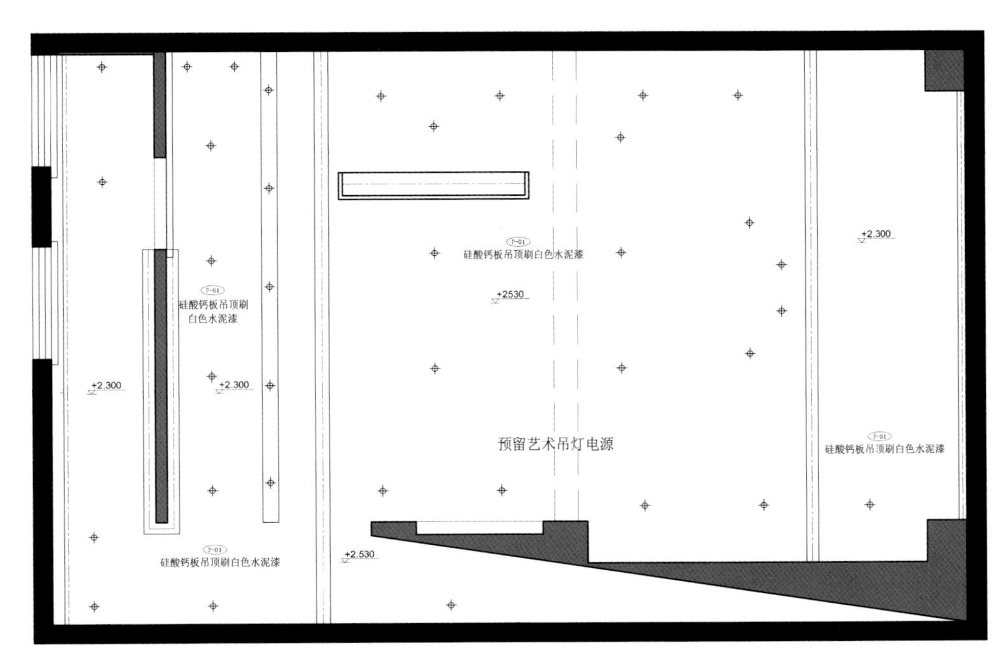

Ceiling Plan / 吊顶平面图

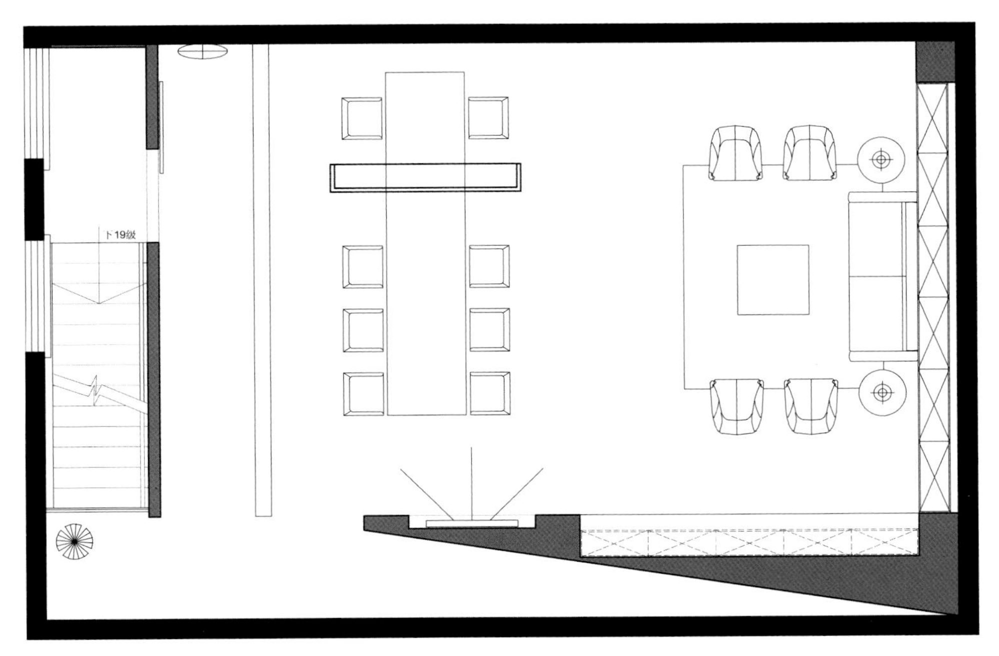

Second Floor Plan / 二层平面图

Cardillo Reception Center

石艺汇接待中心

Design Company: Daohe Design
Designer: Gao Xiong
Participant: Guo Yushu
Area: 200 m^2
Main Materials: Black Titanium, Channel Steel, Composite Stone, Square Tube, Wooden Surface Paint, Maple Veneer, Diatom Mud

设计公司：道和设计
项目设计：高雄
参与设计：郭予书
项目面积：200 m^2
主要材料：黑钛、槽钢、复合石材、方管、木面烤漆、枫木饰面板、硅藻泥

People are longing for a kind of relax and simple space, and to get rid of complex and noisy in this increasingly busy life.

It's a marble composite panels exhibition hall. The designer introduced the characteristics and property of products to the design thought, making the product display more generalized and life-styled. The black titanium, channel steel, square tube and wooden surface paint are used to make the door head and display wall more tridimensional, thus the product itself will be expressed more delicately and bring clean, hale and hearty feelings. The unique landscape design of bridge house water system, makes a rustic experience to people. A sink-style square reception area is used to make a sense of hierarchy. Natural light is introduced through the circular skylight, which goes well with the sink-style square area, presenting a Chinese proverb of "no compasses and carpenter's square, no circle and tetragonum" which can be also explained that "Nothing can be accomplished without norms or standards".

在这日趋繁忙的生活中，人们渴望得到一种能彻底放松、以简洁和纯净来调节精神状态的空间。人们在互补意识的支配下，产生了亟欲摆脱繁琐、复杂、追求简单和自然的心理。

本方案为大理石复合板展厅，设计师从其产品的性能、特点出发了引入设计思维，为使产品展示全面化并贴近生活，“简约”便成为了设计的中心词汇。设计应用了“面”的区分法，利用黑钛、槽钢、方管、木面烤漆等，将门头及大厅展示墙立体化，即抛开原始的墙面展示，也不采用古板的层架展示，从而更精致地表达产品本身，给人以干净、硬朗的感觉；独具匠心的过桥入户水系景观设计，打造“小桥流水人家”的独特体验。本案应用了下沉式的方形接待区，令空间富有层次感。圆形天井引入自然光线，与下沉位置相呼应，“无规矩不成方圆”亦是本案设计的精髓所在。

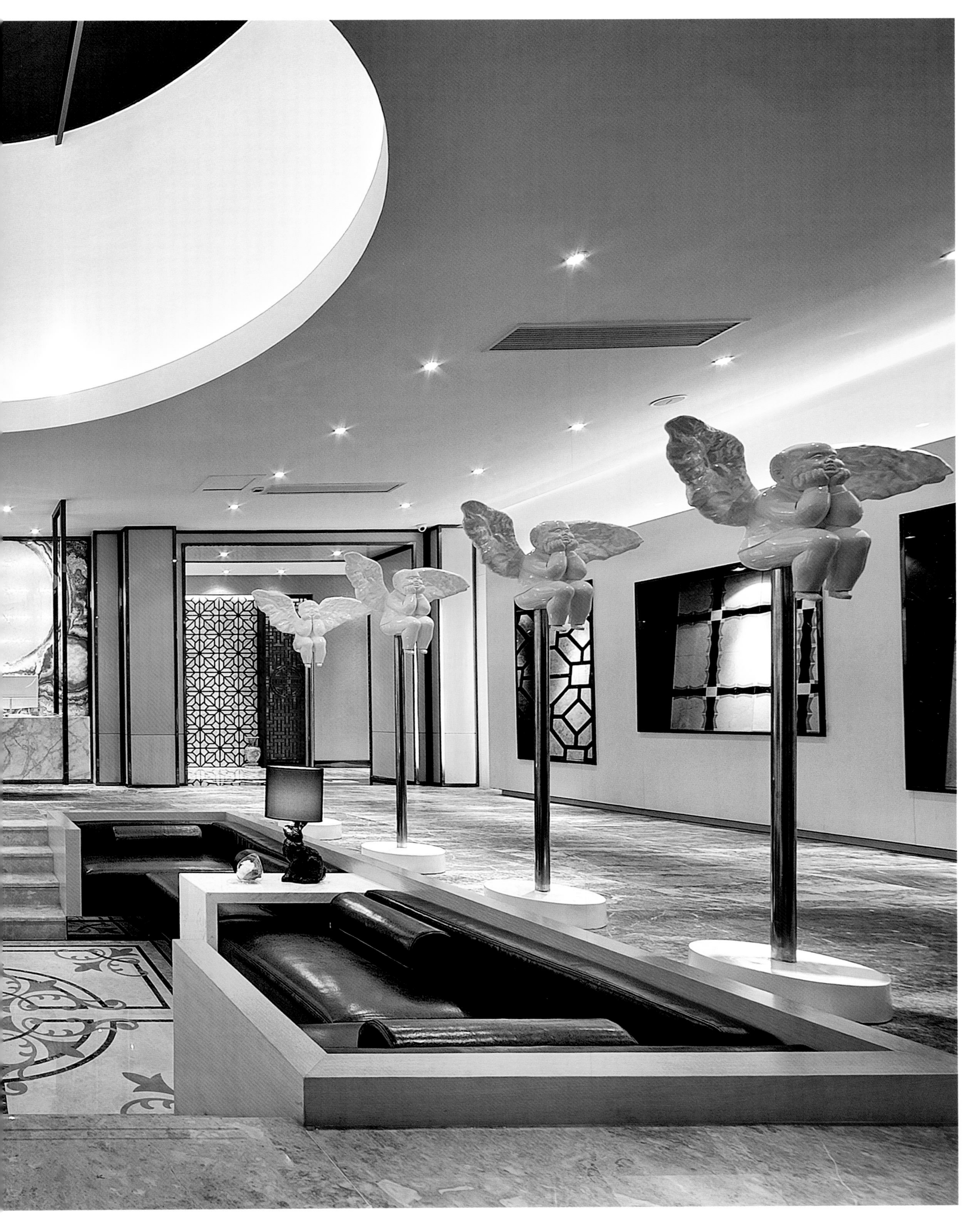

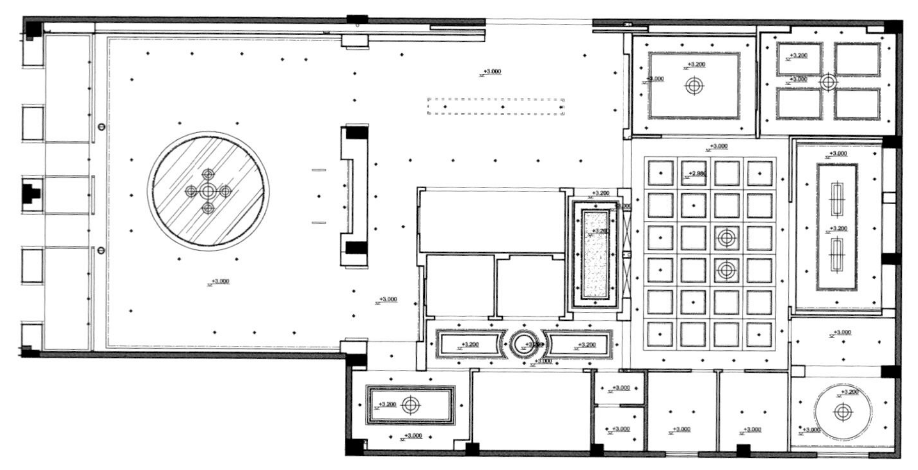

Site Plan / 平面图

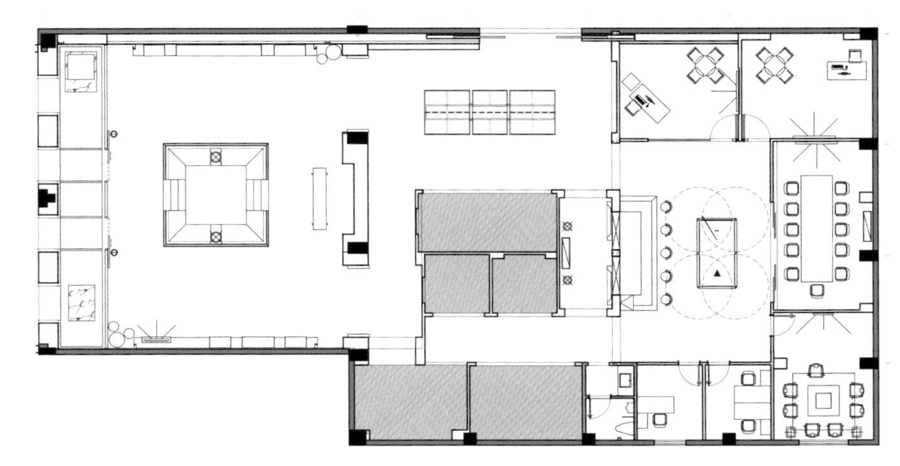

Site Plan / 平面图

CARDILLO

星牌 STAR
星牌 STAR

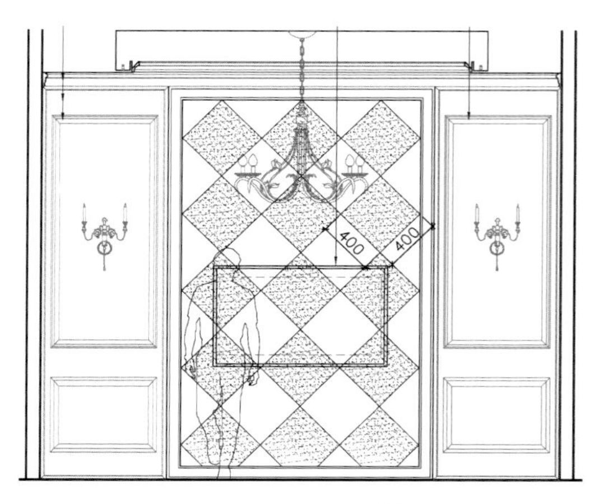

Elevation Drawing / 立面图

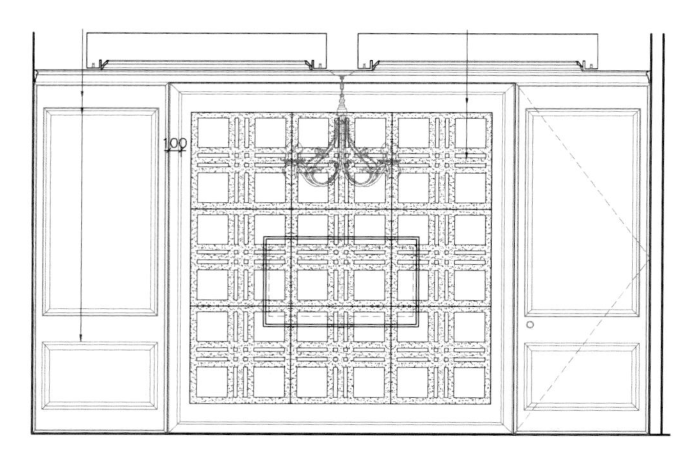

Elevation Drawing / 立面图

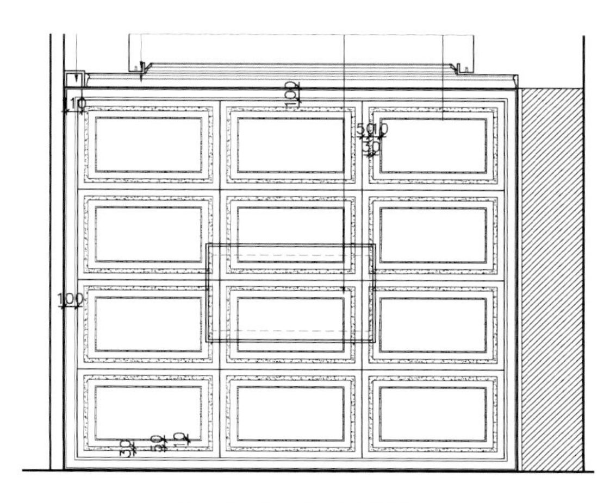

Elevation Drawing / 立面图

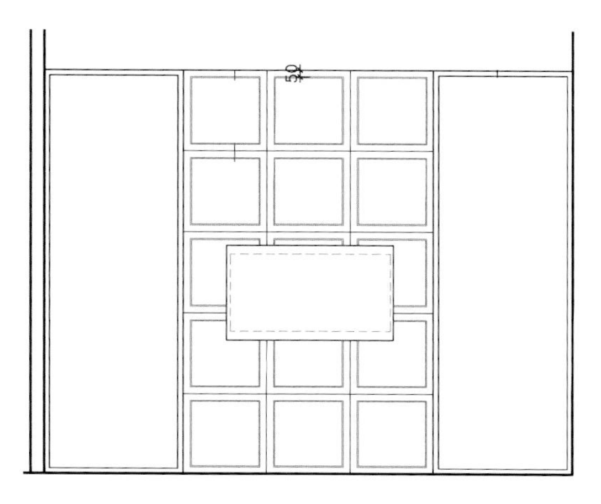

Elevation Drawing / 立面图

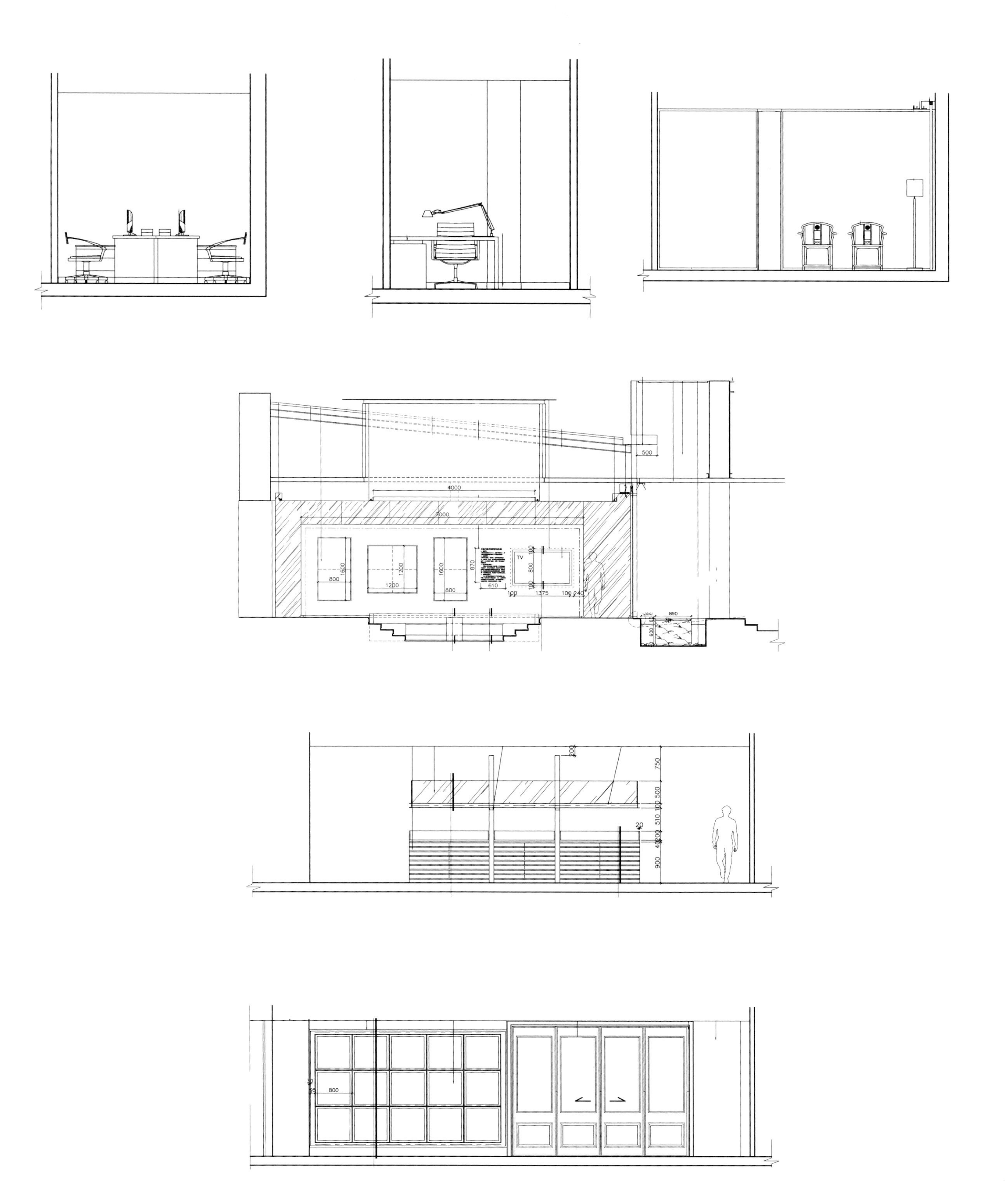

500
4000
7000
1500
800
1200
1200
1600
800
870
610
TV
100
800
100
1375
100 240
890
600
200
750
500
100
510
20
900
800

Chateau Terrior

天瑞酒庄

Design Company: Comeber Design
Designers: Shi Chuanfeng, Xu Na
Photographer: Zhou Yuedong
Area: 180 m^2
Main Materials: Wallpaper, Litchi Surface Marble, Cork

设计公司：宽北设计装饰设计机构
项目设计：施传峰、许娜
项目摄影：周跃东
项目面积：180 m^2
主要材料：墙纸、荔枝面大理石、软木

In Terrior's, people could taste fantastic wine from France, Italy and Portugal in romantic and noble surrounding. The design of building combined with various styles and will be a transmission of wine culture, makes people's dreaming life to be realized.

The chateau is divided into two independent floors, but the architectural structure is designed convenient in communication. The different areas of the space have the advantages of functions and visual enjoyment, which is showed by changing colors, light and materials. This case breaks out of normal design concept. Wine culture is deduced by relative elements. The entrance of the first floor is decorated by cork-shaped columns. At a side of washbasin, there is a oak barrel-shaped table which displayed with decorations and wineglass. The background wall decorated with cork curtain brings new visual experience. When the light reflected on the curtain, it seems like stars shining on the wall. When the wind swept through, it seems like treasure behind the swaying curtain.

The colors and light of space also affects the mood of the visitors besides novel materials. It's incredible that visitors will have a feeling of getting rid of noisy and impetuousness from real life. Designer organized the point light and floodlight ingeniously, and enriched aesthetic sense of space by using unique lighting. As the light falls on the void and solid grating, a fancy dreaming space is emerging loomingly. And as people's footsteps approaching, the ethereal dream will be clear. A different aesthetic scene was showed

produced by seeing from different angles of view. Besides soft characteristic, the light could also build a mysterious effect. This is a kind of variations on the design style, and also is a unexpected simplify of design language. People will find it very attractive in each aspects.

品法国红酒的浓香，赏意大利红酒的色泽，享葡萄牙红酒的甘美，在浓情画意之中品尝真醇佳酿，浪漫、尊贵随之弥漫。在天瑞酒庄里，几乎每一个角落均可视为是对葡萄酒文化的传承与演绎。它的空间情趣与节奏风格融合了多样的风情与文化，使得隐藏于都市人心中关于精致生活的梦想得以实现。

酒庄分为上下两层，它们之间彼此独立，却又不乏交流的可能。空间中的各个区域在满足各自功能的基础上，用色彩、光影、材质的变化引导着人们的视觉感受。这似乎在改变着我们对时尚装修概念的定性思维，设计师充分利用与红酒相关的元素尽情地演绎了多元的红酒文化。一楼入口的锥形柱做成由夸张变形的大“橡木塞”重叠而成的形状，它既是纯粹的装饰片段，又是一种时尚的演绎。洗手台旁置放装饰品与酒杯的“高几”，竟是一个古朴的橡木酒桶，它毫不隐晦其率真与坦诚的面孔。而用软木塞串成的帘子则成为了一面背景墙，为我们带来新鲜的视觉体验。当射光打在上面时，仿如满墙的灿烂繁星，远观则又像折射着光的瀑布，当一阵风吹过，似乎便能看到晃动着的帘子后若隐若现的宝藏。

除材料的新奇外，空间里色彩与灯光的设计也牵动着来访者的心情。置身其中，会有一种奇妙的感觉，仿佛刚步出现实的喧闹，便在这个暖色调的空间里渐渐褪去那份浮躁。设计师匠心独运地将点光源与泛光源进行有机地组合，并用独特的灯具造型来丰富空间的美感。当似虚而实的光影透过格栅、屏风铺洒在四周，影影绰绰地构筑起一方新奇的空间，仿如梦中。而随着人们脚步的临近，飘渺的梦境也一点点地清晰起来。因为视角的不同而产生这种不确定的美感，使得灯光除了赋予空间柔和的特质外，还营造出些许神秘的效果。这是设计风格上的一种变调，亦是设计语言中一种出乎意料的洗练，让人们发现这里的每一个层次皆有动人之处。

天瑞酒庄

葡萄酒世界："旧世界"与"新世界"
Vieux Papes

PKO Bank Reception Center, Polski

波兰 PKO 银行接待中心

Designy Company: White Cat Studio
Designer: Robert Majkut
Photographer: Szymon Polanski

设计公司：白猫工作室
项目设计：Robert Majkut
项目摄影：Szymon Polanski

The starting point for this project was Bank's corporate identity developed by White Cat Studio, above all the modernized logo of PKO Bank Polski in its elegant color for Private Banking sector – black, white and gold, which created a set of basic colors for the interior. Other graphical elements were also inspiring, decorative motif consisting of a grid of elegant, sinusoidal lines, consistently applied in the graphic design for the PKO Bank Polski Private Banking.

The theme of delicate grid was treated in a very innovative way, a bi-dimensional pattern was completely transformed by introducing an additional dimension: it was made spatial by being projected on the tri-dimensional model of the interior. To achieve this effect special software for parametric design was used, which allowed for the creation of a complex and ordered structure formed as a transformation of the subtle grid of lines converging in one abstract point. This complicated geometry setting has become a model to be filled with interior design solutions.

As a composition it is coherent and complete, being a real challenge at the stage of realization. The final effect is a balance between purely aesthetic and a mathematical

Bankowość Prywatna

Bankowość Prywatna

order, accepted by the designer, yet generated by software that was treated not only as a tool, but also as a creative factor. It is a kind of metaphorical code which presents the character of the uniqueness and its services, being a mix of mathematical analyses and human element of experience.

本案是白猫工作室为树立银行的企业形象而设计的，私人银行部采用黑、白、金三色作为室内装饰的基本色调，并配以色彩典雅、极具现代感的波兰银行 Logo。其他平面设计元素也同样独具匠心，装饰主题融入了优雅的网格和曲线，PKO 波兰银行的私人银行部也采用了这样的设计风格。

精巧的网格结构经过创意加工，加之第三维的融入，使这个二维模式得到了充分的演绎，体现了室内设计的空间感。设计师用一个抽象原点延伸出来的线条，构建成精致的网格结构，并将此应用于室内软装设计，形成复杂、有序的结构，最终达到体现空间感的目的。本案中复杂几何建构的设置，成了室内设计方案的典范。

图纸上的设计稿完美而清晰，但设计师如何将它变为现实，是一件极具挑战性的事情。需要将纯粹的美感和绝对的秩序完美平衡。室内软装设计不仅仅是一种装饰工具，更是空间设计中的创造性元素，成为一段隐喻的密码，体现了银行机构的独特性和服务性——精确的数学分析与人性化的服务体验。

MOET Chandon Marquee

酩悦品酒大厅

Design Company: LAVA, Gloss Creative
Designers: Chris Bosse, Amanda Henderson
Photographers: Dianna Snape, Marcel Aucar
Area: 100 m^2
Main Material: Lycra

设计公司：LAVA、Gloss Creative
项目设计：克里斯·博斯、阿曼达·亨德森
项目摄影：Dianna Snape、Marcel Aucar
项目面积：100 m^2
主要材料：莱卡

Chris Bosse, together with Amanda Henderson from Gloss Creative, designed the MOET Chandon Marquee for the Melbourne Cup 2005, the biggest annual horseracing event in Australia.The architects used the latest digital technologies from concept sketch to realisation, to create a sparkling and surreal atmosphere in the name of "Bubbleism". Through the use of daylight and a Lycra material that is digitally patterned and custom tailored for the space, which was described as an "avant garde environment not of this earth".

Structure and Space

The project renounces on the application of a structure in the traditional sense. Instead, the space is filled with a three-dimensional lightweight sculpture, solely based on minimal surface tension, freely stretching between wall, ceiling and floor.

MOËT
MOËT
MOËT
MOËT
MOËT

MOËT

克里斯·博斯与阿曼达·亨德森设计的酩悦品酒大厅，是为澳大利亚2005年最大型的年度赛马活动“墨尔本”杯而设计的。从概念草图到建成，建筑师使用了最新的数字技术，创建了一个闪闪发光的，超现实的建筑——Bubbleism。通过日光和莱卡材料的使用，形成了一个数字化的，可自定义的空间，此间只应天上有，人间能得几回见。

结构和空间

这个项目放弃了传统意义上的结构。它在空间中采用了三维轻量级雕塑，在最小的表面张力作用下使墙壁、天花板以及地板之间形成自由伸展。

Turkish Airlines CIP Lounge

土耳其航空 CIP 休息室

Designer: Autoban
Participants: Gokhan Uzun, Sedef Gokce
Photographer: Bülent Özgören
Client: Turkish Airlines-Do&Co

项目设计：Autoban
参与设计：Gokhan Uzun、Sedef Gokce
项目摄影：Bülent Özgören
项目开发：Turkish Airlines-Do&Co

Turkey's national airline, Turkish Airlines CIP Lounge has been opened at the Ataturk Airport International Departures in Istanbul. Designed by Autoban's architectural standpoint, the lounge is spread over 3,000 m^2, with a daily capacity of 2,000 people.

Taking into account the primary purpose of the space is to transmit the "Contemporary Turkey Experience" to Turkish Airlines passengers. The design concept is based on the idea of a second shell within the existing shell of the airport hall. The main structure, established by making full use of the traditional architectural arcade system, formed a globular. These plain spheres create interior combinations by dividing the place into sections, allowing transitions between them.

CIP Lounge's each module undertakes a different function. Services like resting rooms, restaurant, tea garden, library, movie theater and so on provide passengers a chance to experience each and every different interior separately. Spatial organization, which gives the sense of discovering while moving through the lounge; is the natural outcome of the desired architectural layer, which places the mentioned "experience" forward.

Black channels, located in the merging points of modules, are designed to allow mechanical and electrical systems. These channels are among the vital details that bring visual balance to modules which are produced in accordance with the standard of industrial design.

INTERNATIONAL DEPARTURES
INTERNATIONAL DEPARTURES

With Autoban's approach to experience the space at the CIP Lounge, Turkish Airlines, is aiming to carry the airport experience to a higher level.

土耳其国家航空公司的土耳其航空 CIP 休息室位于伊斯坦布尔的阿塔图尔克国际机场。该 CIP 休息室共 3 000 多平方米，每天可容纳 2 000 人，由 Autoban 设计。

考虑到该空间的主要目的是同土耳其航空公司的乘客展示“土耳其的现代体验”。该项目的设计理念是基于机场大厅内部现有的外形基础而设计的第二个外形壳。建筑的主要结构充分利用了传统建筑的拱廊系统，形成球状形式。这些平地上的球体通过了区块的划分形成空间联合体，它们之间可以相互转换。

CIP 休息室的每个模块都有不同的功能。如休息室、餐厅、茶室、图书馆、电影院等等，均让旅客体验到每一个空间的特别。当穿过休息室时，不同的空间组织都会产生一种新的感觉，这是展现“土耳其现代体验”到理想建筑的自然结果。

黑色通道，位于各模块交汇处，再此可安装机械和电气系统。这些通道是至关重要的细节，它们严格按照工业标准设计、生产，实现各模块的视觉平衡。

与 Autoban 一起去感受 CIP 休息室，土耳其国家航空公司的目标是把机场体验带到一个更高的水平。

HIGH-
NEW YORK
WRIGHT

Tiffin Bay

Tiffin 湾

Design Company: Design Spirits Co., Ltd.
Designer: Yuhkichi Kawai
Area: 120 m^2
Location: Kuala Lumpur, Malaysia

设计公司：Design Spirits Co., Ltd.
项目设计：Yuhkichi Kawai
项目面积：120 m^2
项目地点：马来西亚吉隆坡

Tiffin Bay has two faces that are totally different in the day as compared to the night when Tiffin Bay turns into a cafe, lounge and club. This difference attracts appeals especially female customers. The space employs glass screen that customers can enjoy the beautiful view of the Petronas Twin Towers. Customers can use Tiffin Bay to comfortably enjoy some beverages after shopping or dining. The concept of the design is water to suit the big tree designed by David Rockwell, which includes the carpet evoking a pond and the 44 m x 15 m optic curtain wall projecting a waterfall effect casting a beautiful warm light. The boldness fabric pattern, the lighting under the kitchen, colored glass, and the clearness of layered glasses make a new and unique face to the space.

Tiffin 湾有两副完全不同的面孔。与白天相比，晚上的 Tiffin 湾则变成一个咖啡馆、休息室和俱乐部。这种差异吸引了更多的注意力，特别是对女顾客。透过空间的玻璃屏幕，顾户可以享受双子塔美丽的景色。购物或就餐之余，顾客也可在此舒适的享用 Tiffin 湾的一些饮品。本案由大卫·洛克威尔设计，其设计理念是用水元素来搭配各种大树，其中包括一方由地毯引至的小池和一个由 44 mx15 m 的光学幕墙打造的瀑布，使空间笼罩在美而温暖的光的世界。大胆的织物图案，厨房的照明，彩色玻璃和清晰的玻璃分层使空间展现出其新奇和独特的一面。

Martian Embassy

火星大使馆

Design Company: LAVA
Designers: Chris Bosse, Tobias Wallisser, Alexander Rieck
Photographers: Brett Boardman, Peter Murphy
Area: 150 m²

设计公司：LAVA
项目设计：Chris Bosse、Tobias Wallisser、Alexander Rieck
项目摄影：Brett Boardman、Peter Murphy
项目面积：150 m²

A Martian Embassy was designed by LAVA as an immersive space of oscillating plywood ribs brought to life by red planet light and sound projections.

LAVA's design, with partners Will O'Rourke and The Glue Society, is for the Sydney Story Factory, a not-for-profit creative writing center for young people in Redfern, Sydney.

The design is a fusion of a whale, a rocket and a time tunnel - inspired by the stuff great stories are made of Herman Melvilles *Moby Dick,* H. G. Wells' *Time Machine* and Stanley Kubrich's *2001: A Space* Odyssey. The concept is to awaken creativity in kids, so the design acts as a trigger, firing up the engines of imagination. It's an intergalactic journey, from the embassy, at the street entrance, to the shop full of red planet traveller essentials, to the classroom. By the time kids reach the writing classes they have forgotten they are in "school".

LAVA used a fluid geometry to merge the three program components (embassy, school and shop), and a computer model was sliced and "nested" into buildable components. 1,068 pieces of CNC-cut plywood were put together like a giant puzzle. Using technologies from the yacht and space industry the timber ribs create shelves, seats, benches, storage, counters and displays and continue as strips on the floor. Edged with Martian green, the curvy plywood flows seamlessly so that walls, ceiling and floor,

THE
MARTIAN
EMPIRE

BLACK HOLE
MARTIAN WHISTLE
VAPORISED HUMAN
HUMAN DISGUISE KIT
PUNT HUMANS
MARTIAN CAPE BUDGET VERSION
MELTED ICE

TRINACRIA

MARTIAN EGGS
BUGS

WELCOME!
GLOB
MERCURY
SOFT SERVE
RATTLES
LAUGH
CHUBBY
TURTLE NECK
SPACE DUST
GARGANTUAN
FASTIDIOUS
LAVA
NASA

TRINACRIA

space, structure and ornament, become one element. Mar inspires young imaginations, whilst the sounds and lights of the red planet animate the space.

火星大使馆由 LAVA 设计。该设计通过红色星光和声音监测产生 " 身临其境 " 的空间感。

LAVA 与 Will O'Rourke 和 The Glue Society 合作为位于悉尼雷德芬的一个年轻人的非盈利创意写作中心——悉尼的故事厂而设计。

该设计是鲸鱼、火箭和时间隧道的融合。收到了这些精彩故事的启发——赫尔曼·麦尔维尔的《白鲸》, H . G . 威尔斯的《时间机器》和斯坦利·库布里克的《2001: 太空奥德赛》。该设计概念旨在唤醒孩子们的创造力，使设计成为一个触发器，一个激发想象力的引擎。这是一段星际旅程——从美国大使馆，到街的入口，到处布满了红色星球旅行必需品的商店，再到教室。此刻孩子们早已忘了他们是在“学校”的写作课了。

LAVA 使用流动几何合并三个组件（大使馆、学校和商店)，一个计算机模型和“嵌套”到切片可建造的组件。1 068 块数控切割胶合板被放在一起就像一个巨大的疑团。空间里使用航天工业木材制造的货架、座椅、长椅、贮存器、计数器和显示器，镶有绿色的火星、弯曲的无缝胶合板、墙壁、天花板和地板、空间、结构、装饰成为一个元素。火星激发年轻人的想象力，而声音和红色的灯光激活了行星的空间感。

Zen

禅念

Design Company: Shenzhen Mincillier Design Consultancy Co., Ltd.
Designer: Feng Yang

设计公司：深圳名斯利烨设计顾问有限公司
项目设计：冯洋

This case is a simple Chinese stylish health club near Lixiang Park. The designer present classical flavor by decorating with ancient prints. The tasteful background is made of ornamental engraving brown glass, and the elegant ceiling is decorated with leaf shaping glass. The theme of design is "Buddha", "Zen" and "Tea", which highlight the connotation of its cultivated function.

本案为荔香公园附近的一家养生馆，整体设计以简约中式风格为主，字画、古诗的衬托，彰显古韵，镂花茶镜制作的背景和叶形茶镜吊顶相互映衬，提升整体格调与档次；最后以“佛”“禅”“茶”的主题，贯穿整体，突出其养生、养心的内涵。

车
楚河
禅
车

禅

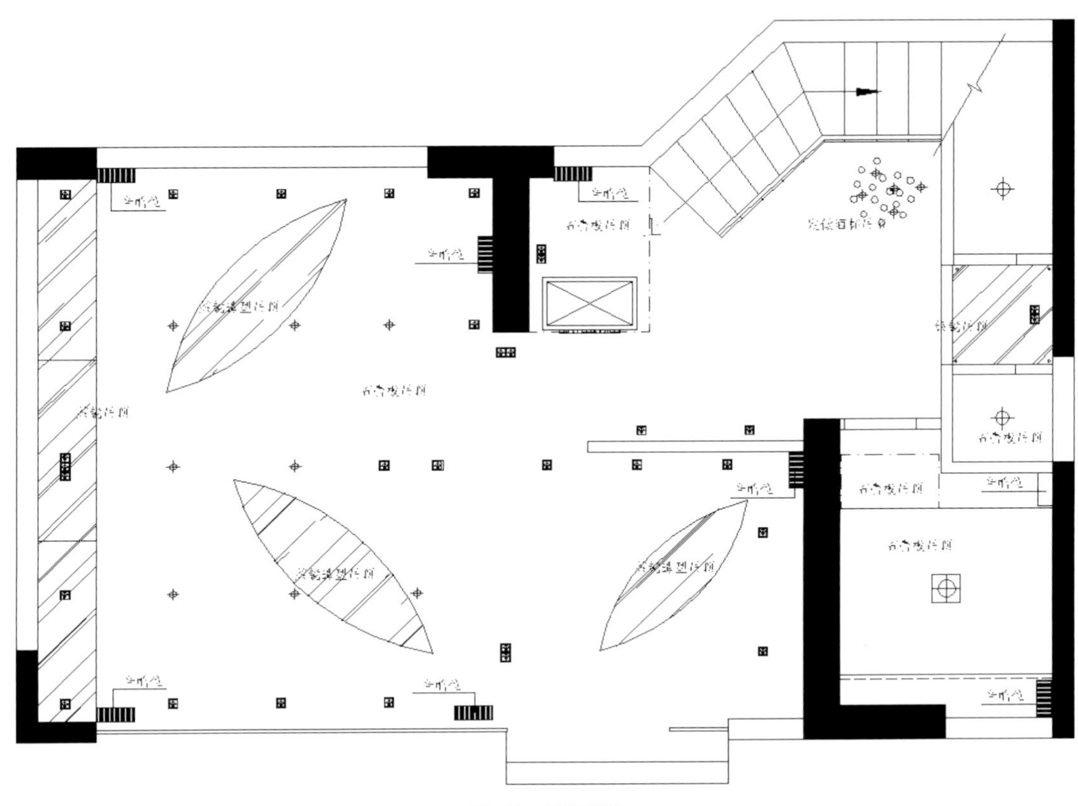

Site Plan / 平面图

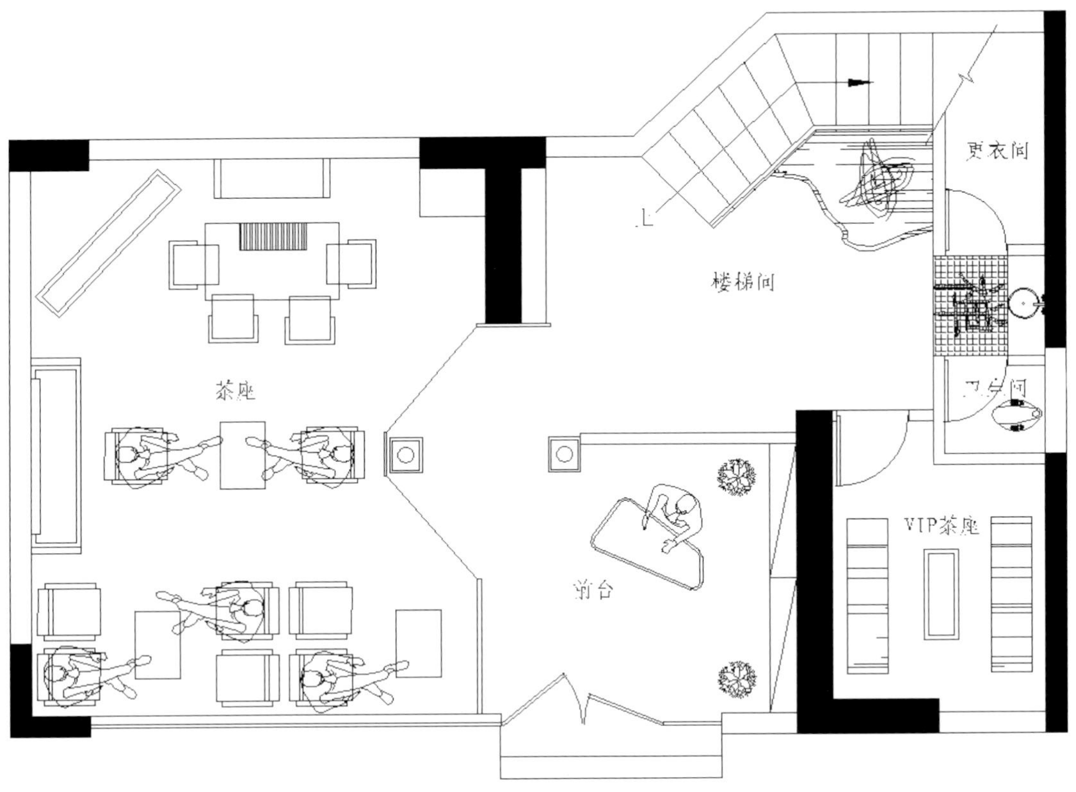

Site Plan / 平面图

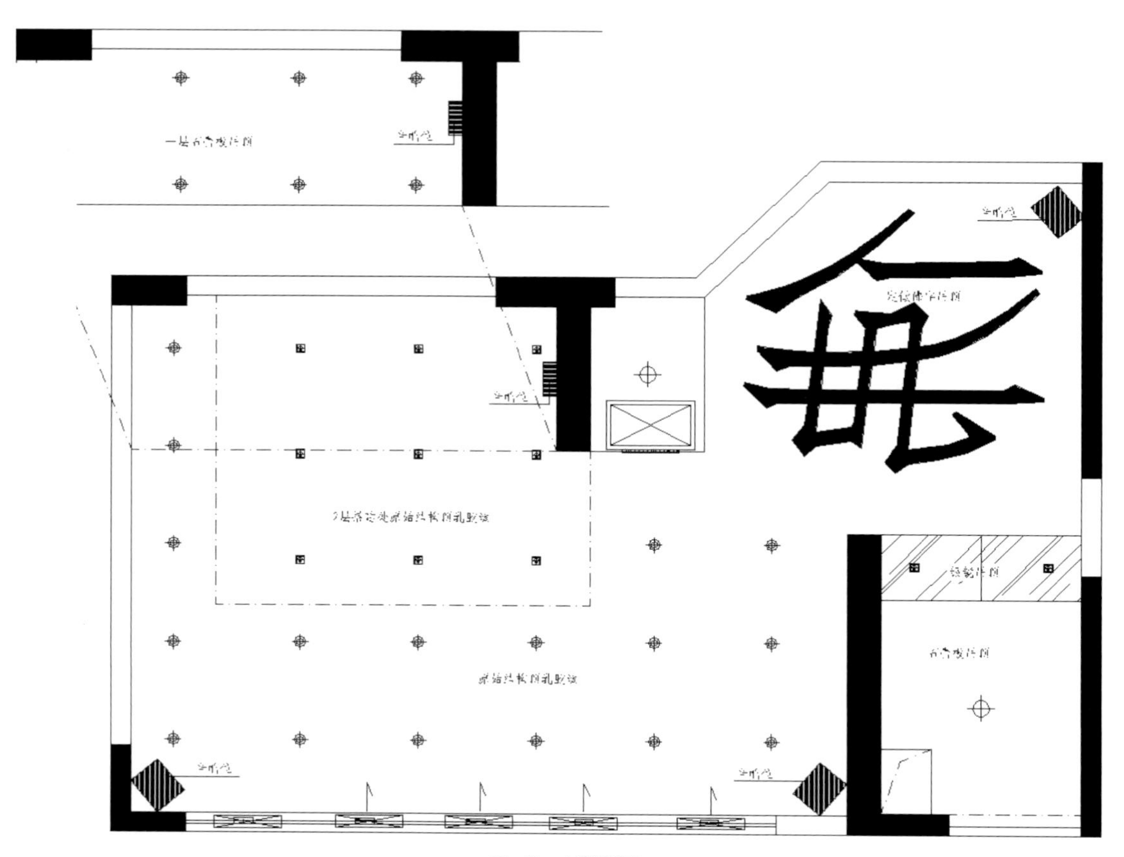

Site Plan / 平面图

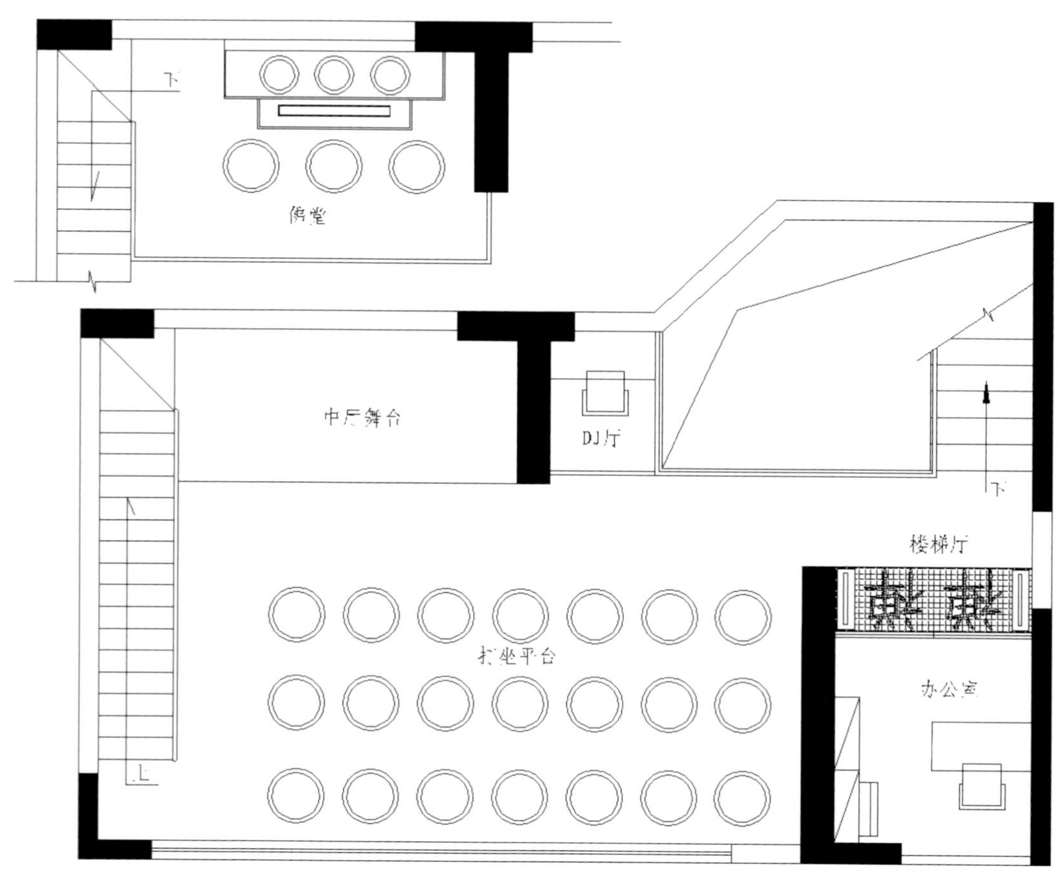

Site Plan / 平面图

禪

Taifu Medical Center

泰福体检中心

Design Company: Hong Wings Design
Area: 1,900 m^2
Main Materials: Wallpaper, Plastic Floor

设计公司：郑州弘文建筑装饰设计有限公司
项目面积：1 900 m^2
主要材料：壁纸、塑胶地板

Creating a context of building is very important in the design besides the endure look of structure and space. A success of design always depends on a interesting and tasteful context. Making a special inventive design also should be the ideal of all designers.

Only healthy people will go to medical center, so the medical center should be a delightful place to create happiness. However, the happiest time of life should be childhood. It's a colorful, free and pure world, where children could be exploring among a forest, or playing under big trees for all day. Designers built a relaxing and interesting space to make happyness, and people will find freedom and dream in here.

EXIT

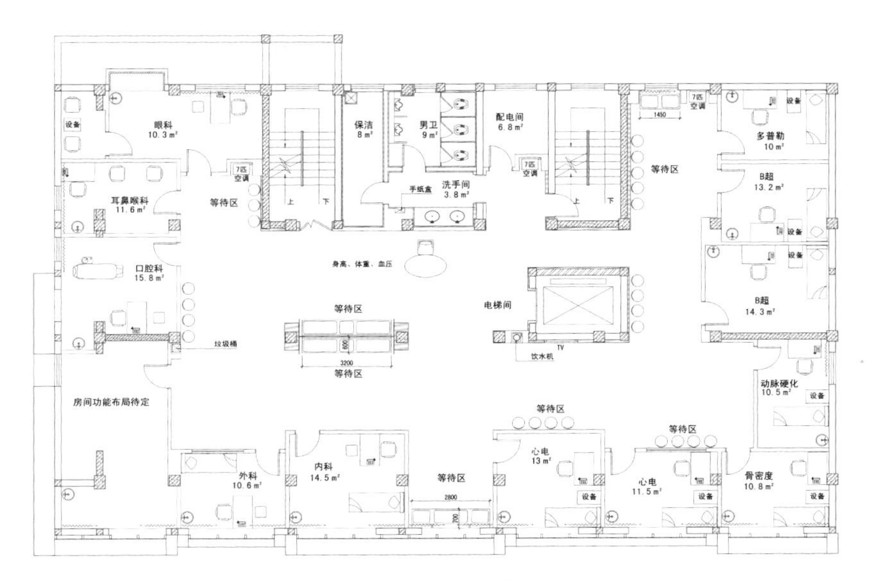

Site Plan / 平面图

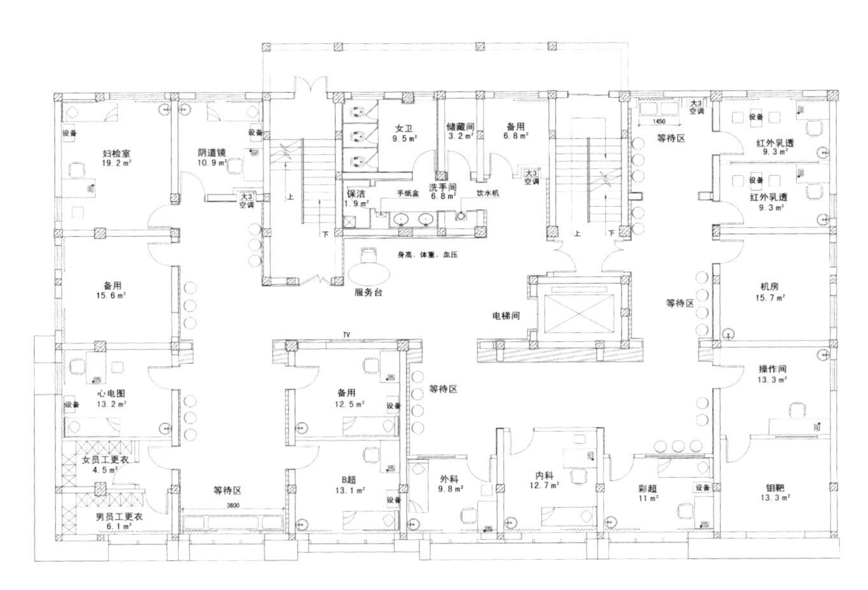

Site Plan / 平面图

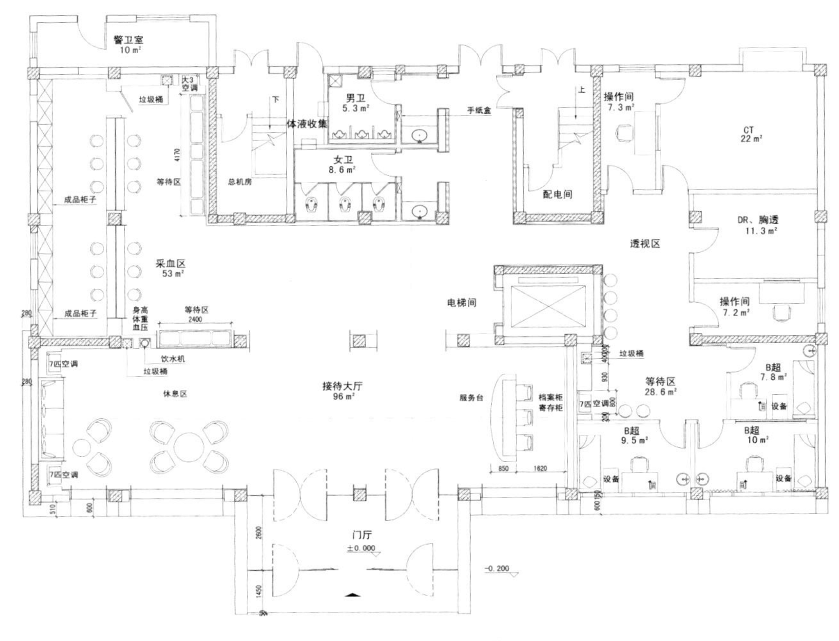

Site Plan / 平面图

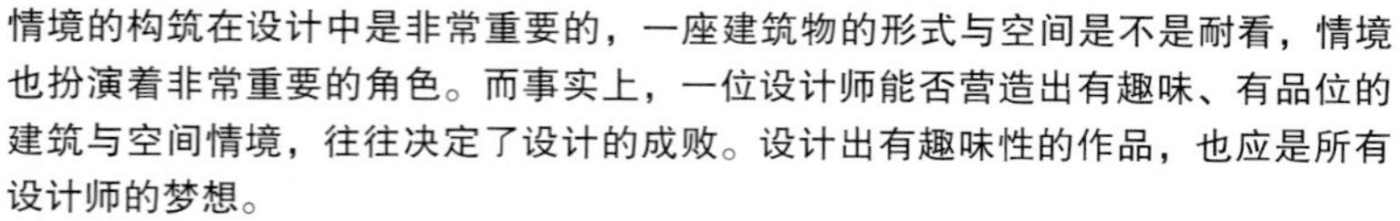

情境的构筑在设计中是非常重要的，一座建筑物的形式与空间是不是耐看，情境也扮演着非常重要的角色。而事实上，一位设计师能否营造出有趣味、有品位的建筑与空间情境，往往决定了设计的成败。设计出有趣味性的作品，也应是所有设计师的梦想。

根据体检中心的性质，只有健康的人才需要体检，因此体检中心就应该是个制造快乐，愉悦身心，成就一次次快乐旅程的地方。而人生最快乐的时光应该是童年，童年是色彩斑斓，无拘无束的，可以自由地在林间戏耍和探索，参差的大树可以成为游戏的世界，一片树叶的叶脉可以让你研究一整天。塑造一个能够自由穿行，制造快乐与趣味的体检空间，在这里你可以找寻到你要的自由和梦。

Garbo Dream Bay Club

嘉宝梦之湾会所

Design Company: Lestyle

设计公司：上海乐尚装饰设计工程有限公司

In this case, "new oriental" design style is introduced into the whole spaces decoration, presenting the aesthetic feeling of order, and advocating a comfortable idea as usual. White pure solid cylinders are used to be partitions of the space. The contrast of materials, the echo of light and shadow makes the space more permeable and open.

The courtyard is the highlight point of the architectural space. Wooden veneer and grainy stone are arranged regularly, and the stable white floor makes people's mind comfortable. The regular arrangement presents a style of simple and neat. The materials used in decoration of VIP area show a cultural atmosphere. The detail lines and frame design highlights the delicate texture of the space. The simple modern style is the fundamental key of this case, introducing oriental elements into the design, to build a modern oriental stylish living environment. Water bar area with open space layout, using natural materials, highlights the natural fashion of new oriental style. It reflects a perfect combination of users' characteristics and space style.

In the soft decoration, stylish element of decoration is introduced into oriental design elements. The mix match of materials and humourous style furniture reflects a provocative sense. Mixed elements soft decoration present a contrast of modern east and new classical. The oriental elements also used in the arrangement of decorations, such as birdcage, glass dragon statues. Blue is the main tone of the whole design, to present a sense of stable and quiet. It also builds a standard visual taste. The design blended into a harmonious whole.

财务室

本案整体装饰风格采用“新东方”的设计风格，并将其融入到整个空间的装饰中，简约中透着秩序的美感。设计师一如既往的崇尚舒适，没有复杂的隔断，散发出不一样的简洁。空间的区隔，白色纯净的立体柱面抽离了干扰人们视觉的几何形图案。质感的对比，光影的呼应，给会所注入了新的空间感受，也使空间更具通透性和宽广的视野。

建筑空间用庭院作为装饰亮点，亲近自然；大体块的木饰面和内敛稳重的木纹石，排列规整，内敛稳重的细化白地面，悄然地抹去人们心灵的拥挤。规整的排列更显大气，增强空间的延续性。空间色调简单、素净，体现了建筑的空间感。VIP 区材质的结合展现出空间的人文气息，而细节线条、边框的设计则凸显空间的细腻质感。VIP 区以现代风格的简洁线条为基调，跳脱现代风格一成不变的空间形式，将东方时尚元素融入空间，打造现代东方风格居住环境 。水吧区则采用开放的空间布局，运用自然的材质转换，凸显新东方的自然、时尚。本案之简约，将空间使用者的特质与空间风格完美结合。

在软装设计上，设计师将东方精髓设计元素融入到装饰主义风格元素中。这不单单是东方元素的简单延承，而是加入了新摩登元素。具有幽默感的家具，装饰材质的大胆混搭，将张扬的装饰主义展现得淋漓尽致。而家具的软装混搭的不同元素，摩登东方与现代新古典的拼撞，无疑是装饰主义最好的表现。在饰品的选用上，东方元素始终贯穿其中，如造型各异的装饰鸟笼，围廊四周陈列极具东方意境的龙生九子琉璃雕像，也寓意着吉祥。整个设计以蓝色为主调，色彩稳重而沉静，也形成统一的视觉审美，设计浑然一体。

Tradies Union Club

Tradies 工会俱乐部

Design Company: Red Design Group
Area: 340 m²

设计公司：Red Design Group
项目面积：340 m²

Tradies and Red Design Group's collaboration have set a new benchmark for club design in Australia. The team at Red Design Group were commissioned to develop a new design and brand that would connect with the local Sutherland Shire community. Tradies CEO Tim McAleer wanted a design that would "keep us relevant and yet make us different from everybody else". Not only did the design achieve the brief but it also brought "the wow" factor.

Red Design Group understood that they not only needed to create a space that was aesthetically pleasing, but that also increased memberships for the client. Since the foyer opening in April , 2012 Tradies has gained an extra 2 000 members; which is a testament to how well thought out design can attract an organisations target market.

The team at Tradies wanted to differentiate themselves from other similar clubs found throughout NSW. They requested a design that would combine with the local community "and if anybody comes to look at what's being achieved here is something absolutely unique" stated Tim McAleer in regards to the outcome.

Tradies
Tradies

Tradies
Tradies
Tradies

$ 551.86
$ 41.45

Tradies
Jackpot
LANTERNS

Jackpot
LANTERNS
$637.91
$20.03
$ 790.60
$ 28.03
Tradies

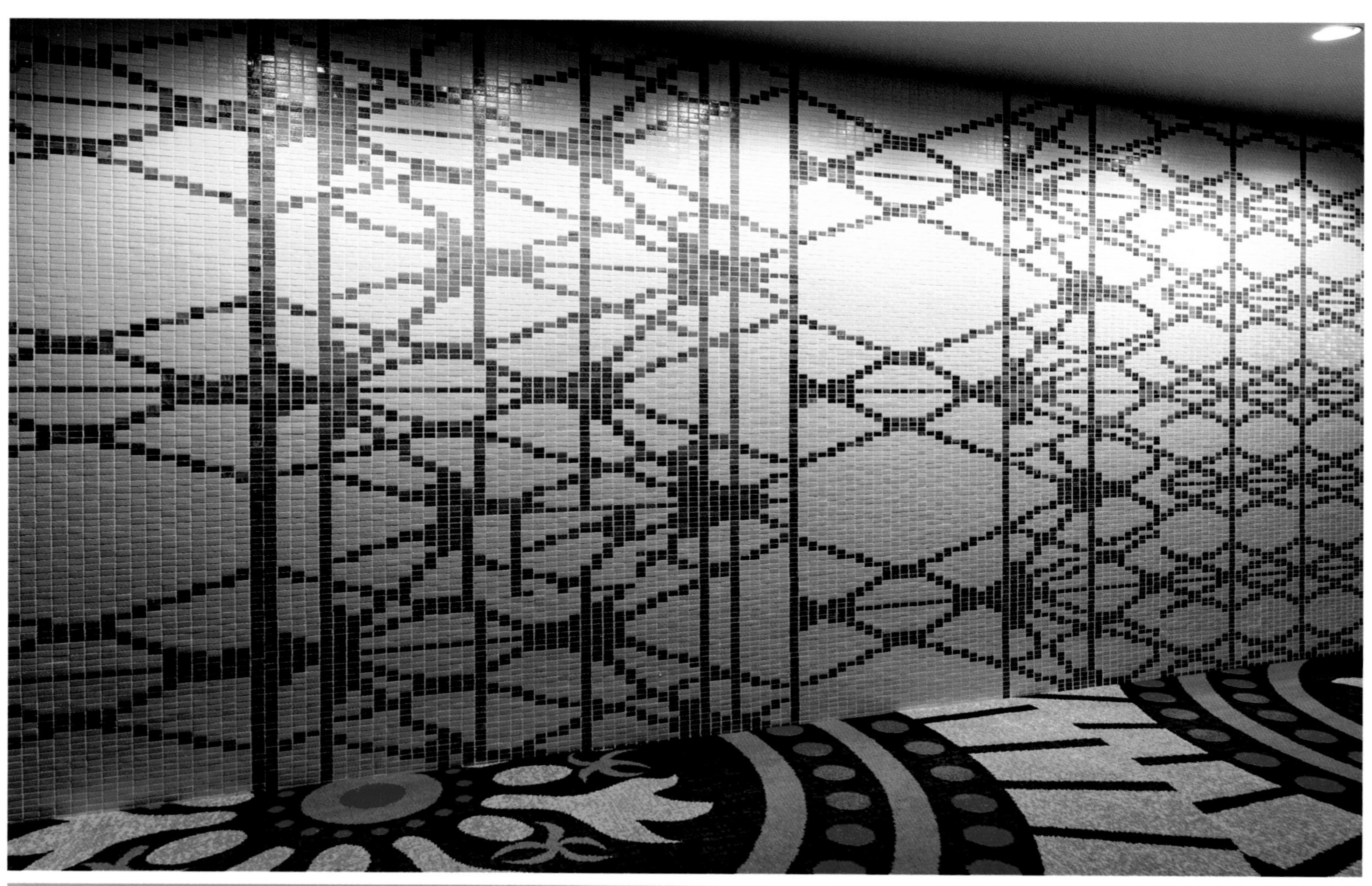

Tradies 和 Red Design Group 合作在澳大利亚建立了一个新的俱乐部。Red Design Group 受委托开发的新设计和品牌，将与本地萨瑟兰郡社区连接。Tradies 首席执行官 Tim McAleer 想要一个能实现与其相关，却又特立独行的设计。设计不仅要简洁而且要达到令人惊叹的效果。

设计公司明白 Tradies 要的不仅仅是一个赏心悦目的空间，更是一个可吸引客户的场所。自 2012 年 4 月开业以来，Tradies 会员新增 2 000 多名，这也恰好证明了一个好的设计有助于市场开发。

Tradies 想使其工会有别于其他类似的、遍布新南威尔士的俱乐部。他们要求设计能与当地社区结合，“人们看到的东西要绝对的独特” Tim McAleer 说。

Binge Billiards Theme Club

宾格台球省体店

Design Company: Daohe Design
Designer: Gao Xiong
Participant: Gao Xianming
Photographer: Zhou Yuedong
Area: 800 m^2
Location: Fuzhou, Fujian
Main Materials: Black Titanium, Ashtree, Black Mirror, Wooden Lattice, Grainy Brick, Mongolia Dark Flamed Stone Plate, Wallpaper, PVC Pipe, Carpet, Greenerwood

设计单位：道和设计
项目设计：高雄
参与设计：高宪铭
项目摄影：周跃东
项目面积：800 m^2
项目地点：福建福州
主要材料：黑钛、水曲柳、黑镜、木质花格、木纹砖、蒙古黑火烧拉槽板、墙纸、PVC 管、地毯、绿可木

Geometry cutting method was using in decoration of the entrance and the pool cue cabinets. The background is decorated with irregular PVC pipes, and with the logo embedded in it, drawing a modern geometrical form. Laminate board is used in the decoration of waiting area's wall to widen the space and to be a partition of room. It's not only satisfied the space requirements, but also enhance the light interaction inside and outside. The personal training area is surrounding by two bar counters. The personal training area is not isolated from others, the design style of wall on the left aisle goes with the model of front desk, and display tables are placed among there.

The style of line design is strong and hard. With the changing light, it presents a modern fashion entertainment club. The decoration of space shows a strong contrasts aesthetic. Even if you are not a billiard fan, you will also want to try out of it.

Site Plan / 平面图

入口的前台以及两侧的寄杆柜大胆的将几何切割法运用其中，背景的 PVC 管烤漆经处理形成一定的落差，再将本案的标志嵌入其中，勾勒现代几何形体。由金刚板斜拼组成墙面的休闲等候区，实现了空间的延伸，并且起到隔断的作用，不仅满足空间需求，也增强了包厢内外的光影互动。私教区由两组可供休闲的吧台合围而成，虽称之为“私教”，却并不拒人千里，左边过道的墙面沿用了前台的木作凹凸切割面造型，将展示柜分散在其中。

阳刚硬朗的线条，贯穿于整个空间。用光影的诉说，将格调定位为现代时尚的娱乐台球会所。整个娱乐场所呈现出一种强烈的对比反差美，即便你不是球友，也难免想跃跃欲试。

XING PAI
星牌

XING PAI
星牌

SPA at Elounda Beach

伊罗达海滩水疗中心

Designer: Davide Macullo
Participant: Makis Lahanas
Photographer: Enrico Cano
Area: 1,400 m^2
Main Materials: Steel Structure, White Paint, Teak, Timber, Glass, Marble

项目设计：Davide Macullo
参与设计：Makis Lahanas
项目摄影：Enrico Cano
项目面积：1 400 m^2
主要材料：钢结构、白漆、柚木、木材、玻璃、大理石

This refurbishment project of the Elounda Beach Resort Hotel penthouse into luxury SPA treats the internal and external spaces as a whole. From the outset, the integration of the striking surrounding landscape has been the most important concern to the design development. The principle intention of the project is to ensure a visual continuity between the internal and external spaces but moreover, to extend the space beyond its physical boundaries and into this remarkable surrounding natural landscape. Visitors' perception of the spaces is heightened and the feeling is one of "floating" above this beautiful Crete setting.

The series of Grecian-like "white boxes" which host the various spa functions afford the structure a spatial harmony and are easily understood and navigated by visitors. The material, color and texture choices create a sense of lightness, freshness and purity expected of a place in which body and soul are pampered and treated.

On arriving to the second floor reception, guests have access to two "paths": water based treatments or the second path—facials, massage and body treatments. From the large lounge area through the glass partition, guests can enjoy uninterrupted views out to the panoramic pool, islets and Aegean Sea.

The interior spaces, characterised by the use of "light" materials such as glass and translucent panels for walls and ceilings, create a calming serene, visual neutrality and are gently animated by the effects of reflection, while the surrounding landscape dominates the space and sheds its colors. The only exception to this palette is the bathtub which is dramatically fitted out with black marble and printed glass.

The "leaves" pattern (inspired by the local landscape) used throughout the scheme, contribute to stretch the perspective throughout an atmosphere of pastel colors and lightness, emphasised by the use of artificial lighting at night.

设计师将伊罗达海滩度假酒店的顶楼翻新成豪华的水疗中心，使得内外部空间连成一个整体。从一开始，开发周边引人注目的景观成为设计关注的重点。项目的原则是确保内外部空间之间的视觉连续性。此外，扩展空间物理边界，使其融入周围的自然景观之中。游客的感知空间得到提升，似乎是“漂浮”在这美丽的克里特岛上。

这一系列的希腊式的“白盒子”，其水疗功能多样化空间结构和谐，但也易于被访客熟悉和通过。材料、颜色和质地的选择给人一种轻松感、新鲜感，在这里身体和灵魂将得到放松和治愈。

到达二楼接待室，客人可以穿过两条“路”：清水治疗和美容、按摩、身体治疗。在大型休息区透过玻璃的隔断，客人可以欣赏到游泳池、小岛和爱琴海全景。

室内空间，以“光”为使用材料，如玻璃和半透明的墙以及天花板面板，创建了一个平静安详、视觉中立和稍具反射效果的动画景象，而周围的景观也影响着空间及其色彩。唯一例外的是浴缸戏剧性地配备有黑色大理石和印花玻璃调色板。

叶子的形式（受当地景观影响）应用于整个项目，有助于通过柔和的色调和光度来延伸其景观，并在夜间加强了人工照明。

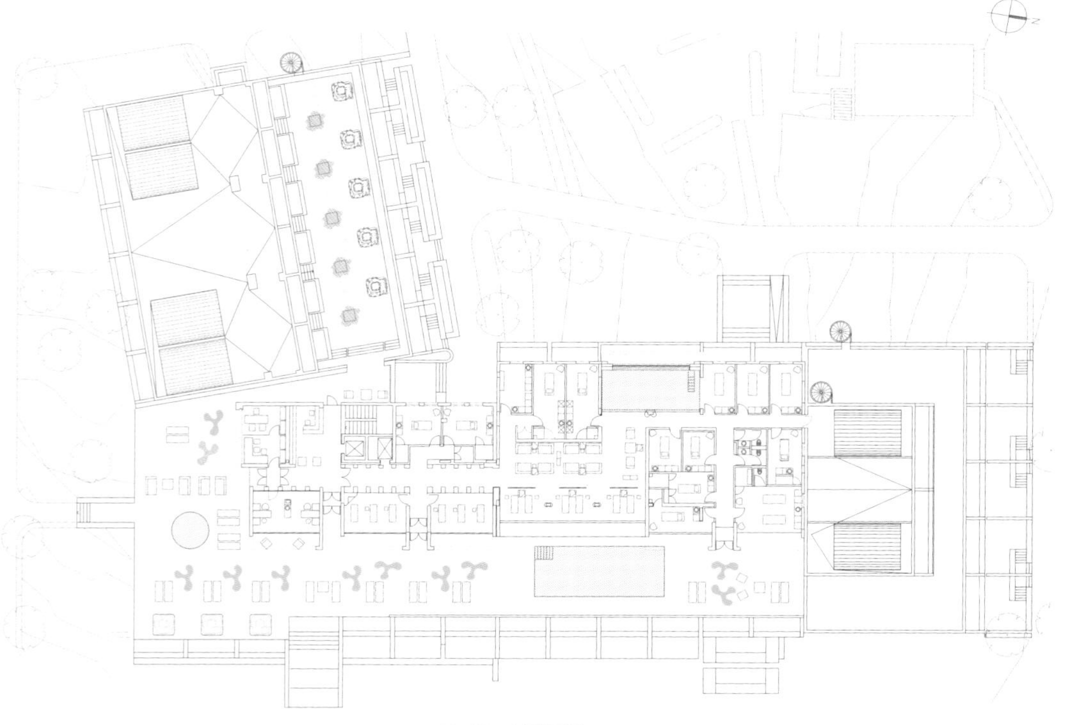

Site Plan / 平面图

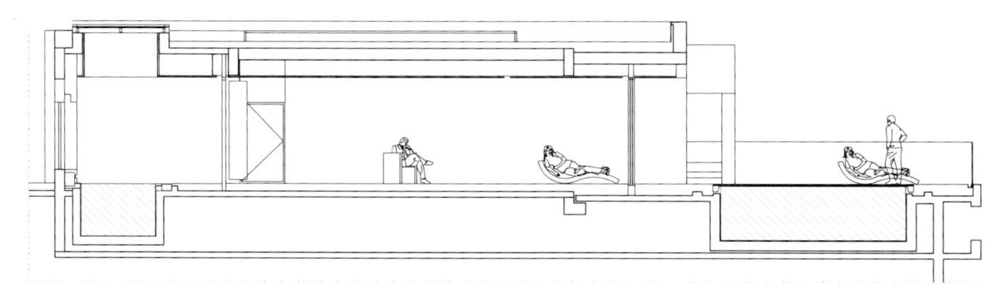

Sectional Drawing / 剖面图

Cleopatra·ME Shine

奇奥柏查 · 美尚城

Design Company: Simon Chong Design Consultants Ltd.
Designer: Zheng Shufen
Soft Decor Designers: Zheng Shufen, Du Heng, Hu Yuan
Area: 12,000 m^2

设计公司：香港郑树芬设计事务所
项目设计：郑树芬
软装设计：郑树芬、杜恒、胡瑗
项目面积：12 000 m^2

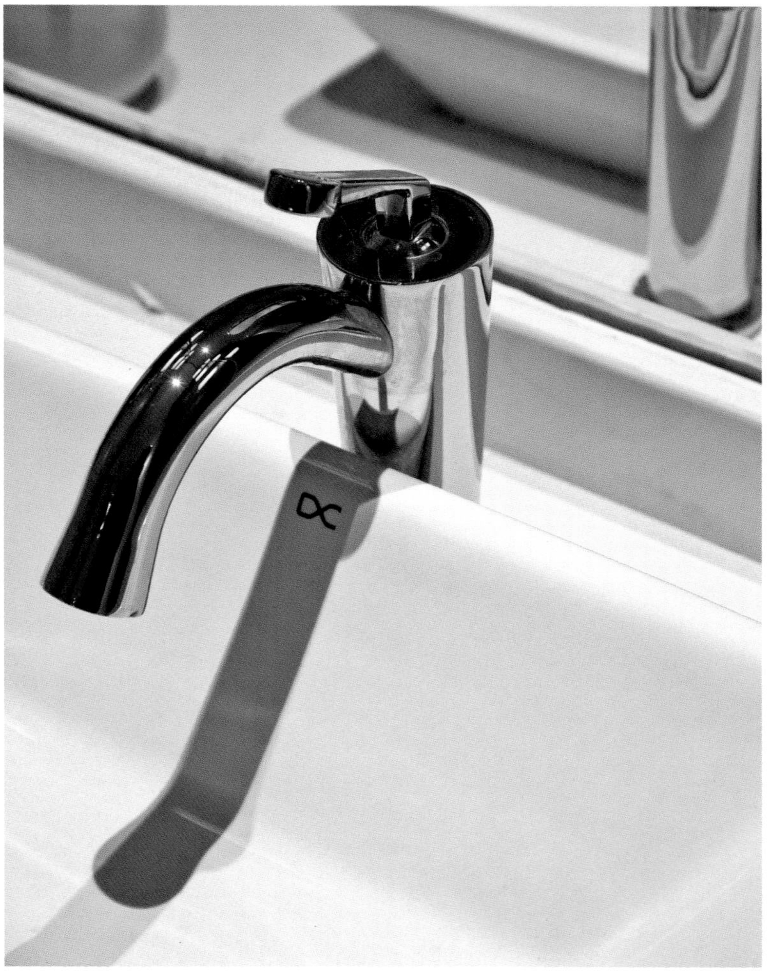

METHODE
BRIGITTE
KETTNER
德國·MBK特色療法
細胞賦活療法
德國·MBK

TOWEL WARMER

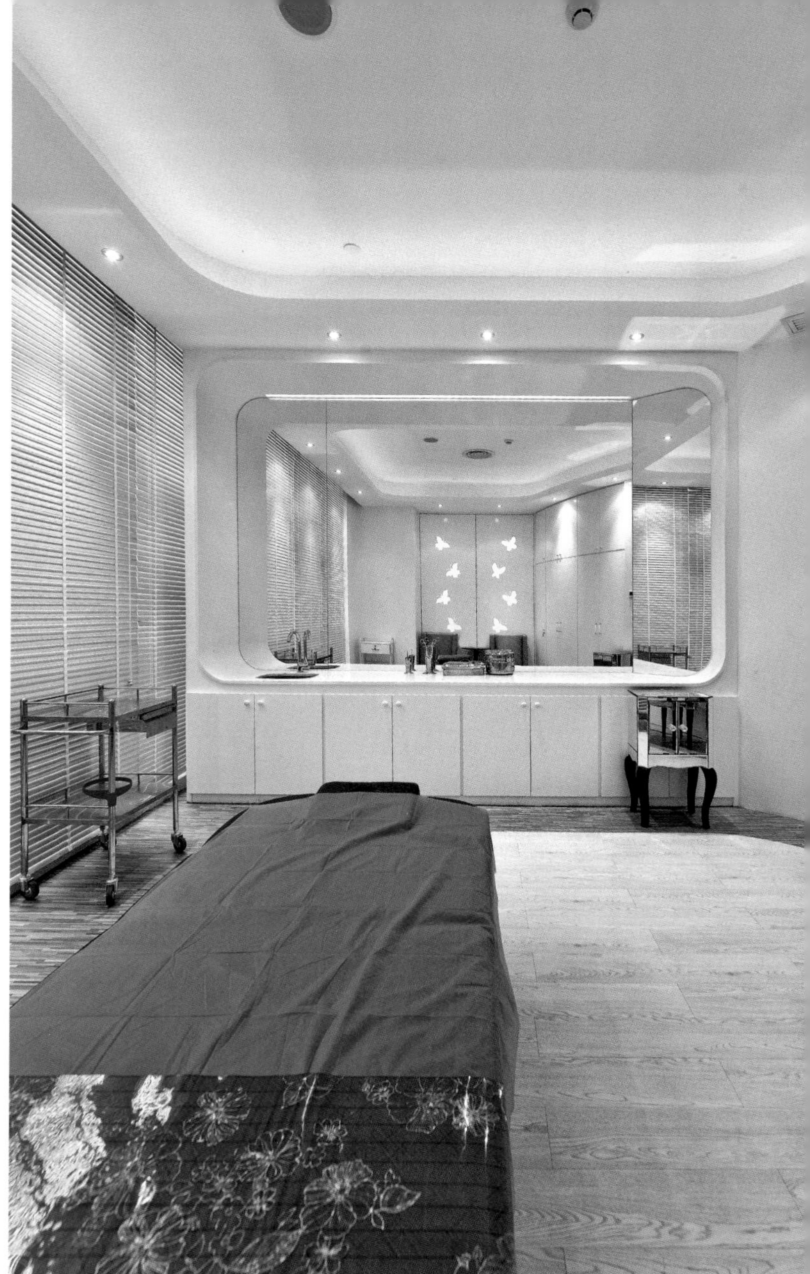

接待区
医生咨询区

健康体检中心
Health Examination Center
接待区

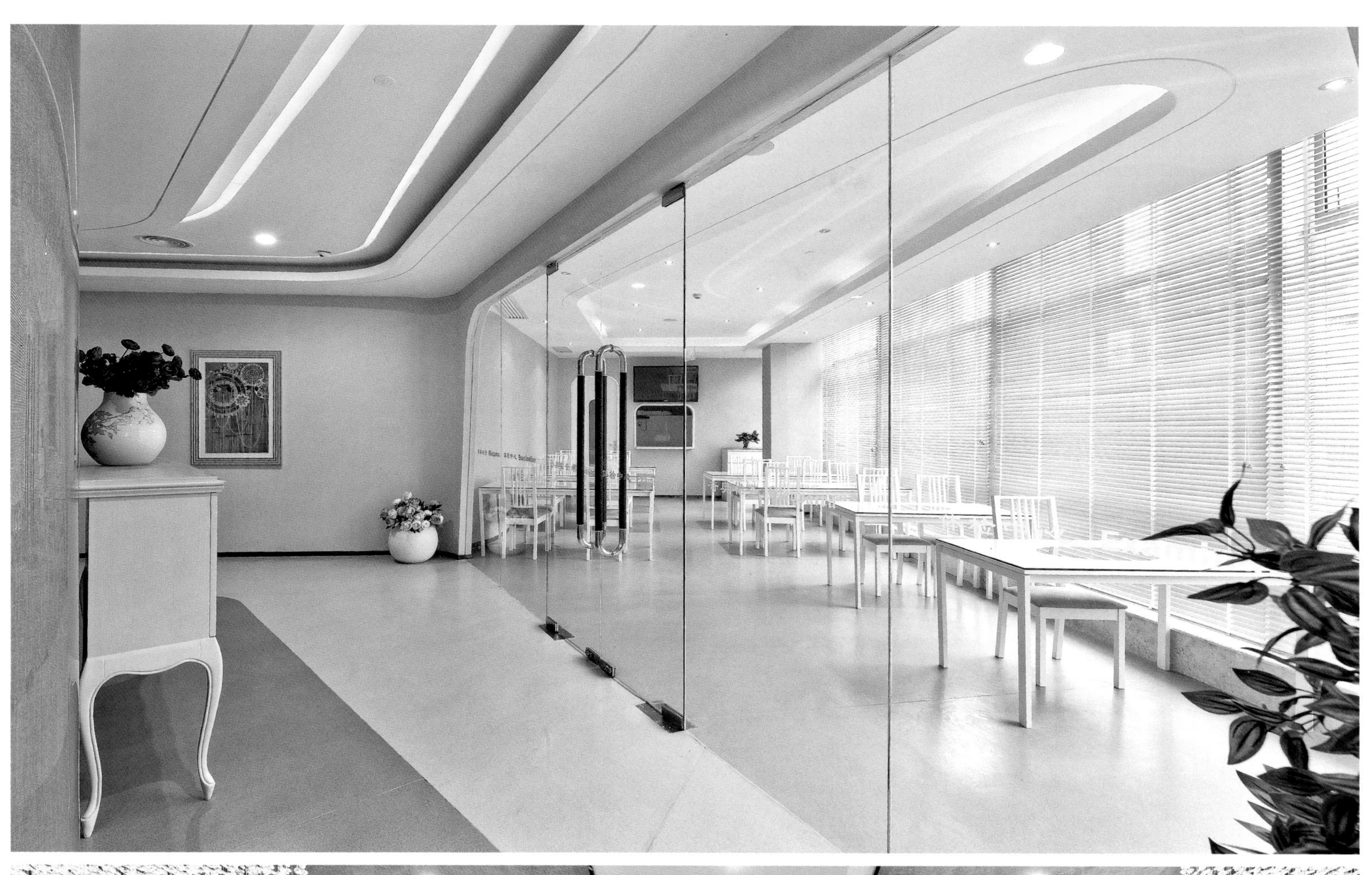

Sales Center of Nanning Ronghe MOCO Community

南宁荣和 MOCO 社区销售中心

Design Company: Eric Tai Design Co., Ltd.

设计公司：戴勇室内设计师事务所

What is MOCO? MOCO explains the real meaning of international life of Move (movable interaction), Own ("I" centered), Cool (giving the special feeling for cities and people), and Original (originally created) – newly born cells of a city. "MOCO" is a "modern group" that is in the leading edge of the period and has the gift of fashionable feeling and seeing clearly fresh things of the life.

The designer applies red, white, black and white bud-shape modeling to express the concept of design. The red color gives us the visual sense for vitality and passion, and the white color reflects pureness of the youth; white acrylic hanging ornaments on the ceiling like a group of pure white life bodies that are sprouting right now, energetic, youthful and fashionable. the dolomite with straight grain on the ground weakens the strangly visual effect which bringed in by red glasses, and leads more subtle changes.

Waved red is the symbol for passion and vitality. Red glass stair railing, red soft wall body

in the negotiation area and red furniture are flame of combustion for the youth. With the companion of deep color wood finishes and black mirror, red and very dynamic gestures emerge continuously, like a red ribbon passing through each field of the space.

The design applies modern and simple geometric shape and the color matching with visual impact to convey vitality and passion, creating a charming space. Youth is not a period of time in your life, but it is the synonym for passion and vitality. Only with youthful mentality, you can stand on the leading edge of period, to become one of the "MOCO" group.

Waved red, waves youth and vitality, waving the creation and passion of "MOCO" group .

何为 MOCO？ MOCO 社区诠释了 Move(移动互动)、Own(以"我"为中心)、Cool(赋予城市与人的特别感受)、Original(原创)国际化生活的真正含义，即一个城市的新生细胞。"MOCO"是走在时代前沿，具有时尚敏锐力与洞察生活新鲜事物天赋的"摩登一族"。

设计师运用红色、白色、黑色及天花白色芽状造型来表达其设计理念，红色给人活力四射的视觉感受，白色体现青春的单纯；天花上密集的白色亚克力吊饰，像一群洁白的、正在发芽的生命体，活力、青春、时尚。地面直纹白云石弱化了红色玻璃带来的强烈视觉效果，产生更多细微的变化。

舞动的红色是热情与活力的象征。红色的玻璃楼梯扶手，红色的洽谈区软包墙体，红色的家具，是青春燃烧的火焰。在深色木饰面及黑镜钢的陪衬下，红色以其富动感的姿态连续出现，像一条红色的飘带贯穿于空间的各个领域。

设计师以现代简洁的几何形态和充满视觉冲击力的色彩搭配来传达活力与激情，营造出一个时尚动感的魅力空间。青春不是生命中的一段时期，而是激情与活力的代名词。只要有年轻的心态，就一样可以走在时代的前沿，成为"MOCO"一族。

舞动的红，舞动着青春与活力，舞动出"MOCO"一族的创意与激情。

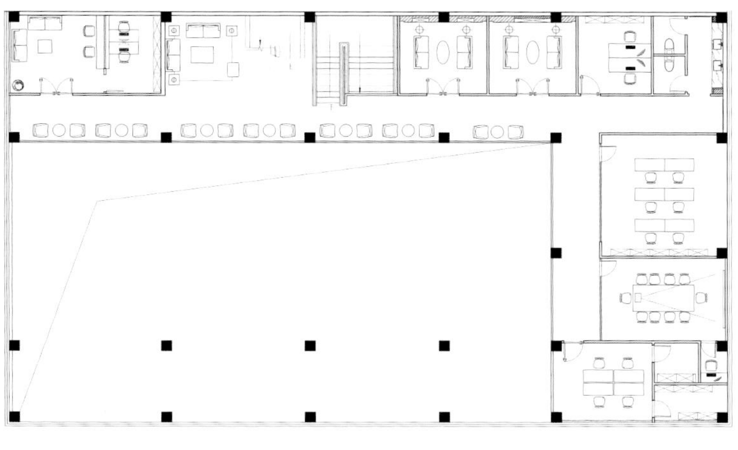
Site Plan / 平面图

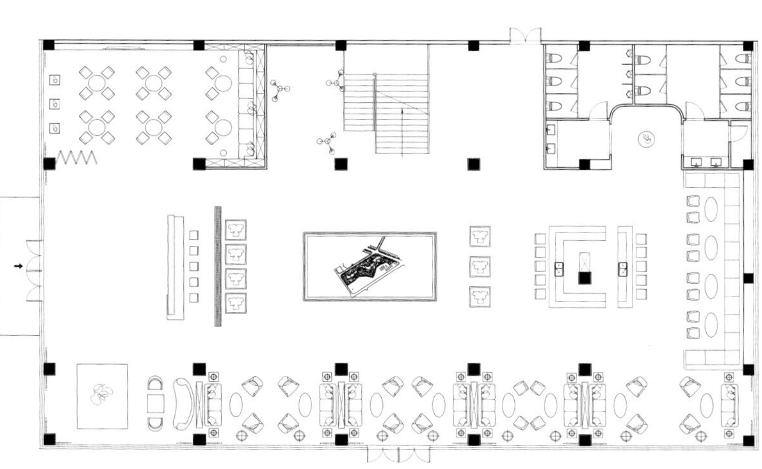
Site Plan / 平面图

Guotai Puhui Reception Center

国泰璞汇接待中心

Design Company: Joy Interior Design Studio
Designer: Joy Chou
Participant: Wu Jialing
Photographer: Lv Guoqi
Area: 658 m^2
Main Materials: Iron, Wooden Lattic, Cement Board, Rosewood, Marble, Glass

设计公司：周易设计工作室
项目设计：周易
参与设计：吴佳玲
项目摄影：吕国企
项目面积：658 m^2
主要材料：铁件、木格栅、水泥板、紫檀木、大理石、玻璃

The design of this case based on integrating the temporary building into the surrounding landscape. Horizontal and vertical lines are used in the design to match the irregular patchwork gray cement plates, and to present the simplicity and delicacy of the appearance of building. The combination of point and ribbonlike light, the green space, and waterscape underline the lightsome beauty of the main building, present vitality from introverted poise with metaphor.

The building is a geometric stretching structural construction with gray steps. The left steel framework embossed agilely and the brown-glass structure present a imagine of amber light box. It makes a contrast with rough cement vividly. The lighting on the stone trails leads visitors to see the scene of borderless pool with point light shining quietly. Two screen walls are built in front of the main building, with windows set on it, to make view more penetrating. Green bamboos are growing under the patio, to maintain privacy and create beautiful landscape.

Opened the bamboo door, visitors could get into the reception center. The bright black floor looks wide and deepen the space, dynamic line design goes with the simple stylish appearance. Striking reception table is the highlight of the large space. Wood and artificial stone structure are used to build the ceiling and foundation of table. The black branches decorated in it present the unique taste of the designer. In addition, the cement is decorated with irregular dark glass, to make the wall looks more stratified, and enable staffs working behind the wall to look after the front desk more convenience.

The models of building under construction are surrounding by grating behind the counter. The diamond cutting white foundation is spotlighted by the light suspended under the ceiling, to present a delicate light effect. The design of the independent negotiating area quite pay attention to guests' privacy, and decorated with oriental elements. The design concept of Suzhou garden —— "borrow scene, circumvent weaknesses" is introduced into this case. Visitors could appreciate bamboo and image of light and shadow from the large glass windows. It provides a safe and private room for a comfortable conversation.

本案的设计基于对临时建筑如何低调地融入周边景观，如何深度演绎极简量体与环境的对应关系。建筑本体以方整的矩形打开，设计用简洁的水平、垂直线条结构，搭配不规则拼接的灰阶水泥板，展现建筑外观的素朴与精致，更结合点状、带状等情境光源，凸显绿地和水景，衬托主建筑的轻盈之美，隐喻内敛中蓄势待发的生命力。

几何延展的灰阶量体，左侧依适当比例浮凸出利落的钢构、茶玻架构，宛若琥珀光盒般，与水泥板的粗犷恰成生动对比。访客驻车后自树篱入口处走上平缓渐升的抿石子步道，随着步道灯光的指引，欣赏沿途相随的无边界水池和倒影于水面的点点光影。主建物前端分别有两道呈 90^0 角的屏风墙，运用细腻的墙体开窗方式，让视觉有条件穿透，使其产生框景般的效果。两墙中间围拱着前方圆形凿孔的天井廊遮，圆孔下方精心栽种的绿竹，取法于自然的绿色建筑概念，在视觉的引导下，透过前后景的巧妙堆栈，兼具维护隐私与美化地景的实质意义。

推开特制竹编大门进入接待中心内部，亮黑色地坪延展出的空间开阔而深邃，动线配置也若简约外观的延伸。醒目的迎宾柜台是巨大的空间光点，由折纸概念而来的立体天棚与柜台基座，分别以木作搭配人造石建构，如同钻石立体切割的形体，在聚光灯的衬托下更显遒劲。横斜其间的黑色树枝，留有设计者独

有的味道。此外，柜台后方的水泥板背墙中错落穿插的黑玻璃，除了让硕大墙体更有层次外，也让墙后办公室内的工作人员，可以透过玻璃孔观察前台动静。

柜台旁有方以大面格栅衬底的角落，用来展示兴建中的建筑模型，同样钻石立体切割的白色基座，透过上方自天花板深处悬垂而下的烟囱式聚光照明，凸显精妙的情境光源效果。独立洽谈区的设计相当注重来客隐私，地面铺设长毛地毯，点缀其上的鼓凳带出微妙的东方人文情愫，“冂”型环绕的沙发同样采用低台处理，而嵌于沙发中央的装置艺术，则利用相互衔接的亚克力棒，结合上下光源，表现烟雾般轻盈的纤柔之美。设计者在这里也特意导入苏州庭园“有景则借、无景则避”的概念，配合横向大面玻璃窗，将窗外灰墙内的绿竹、光影意象吸纳入内，过滤多余街景杂质，也营造一种安全隐密却又无比舒适的洽谈氛围。

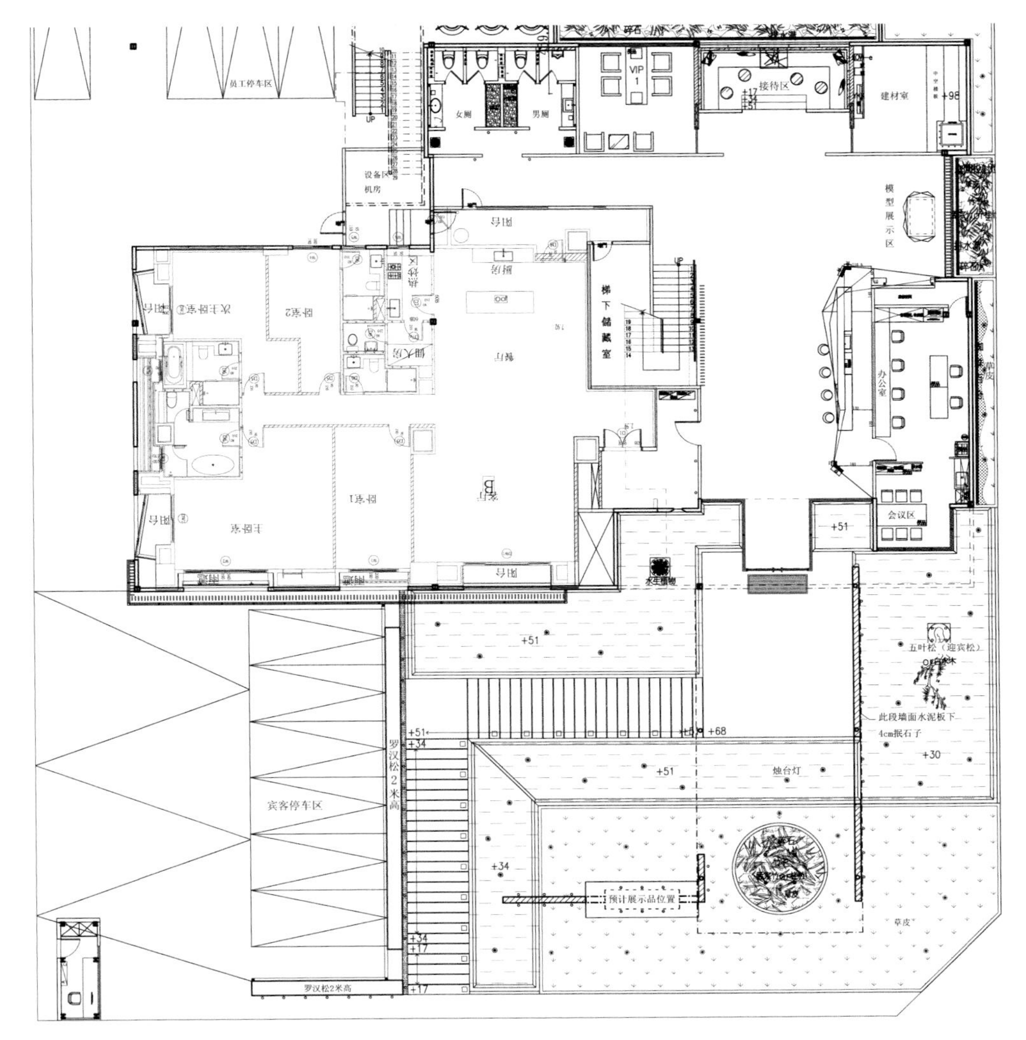

First Floor Plan / 一层平面图

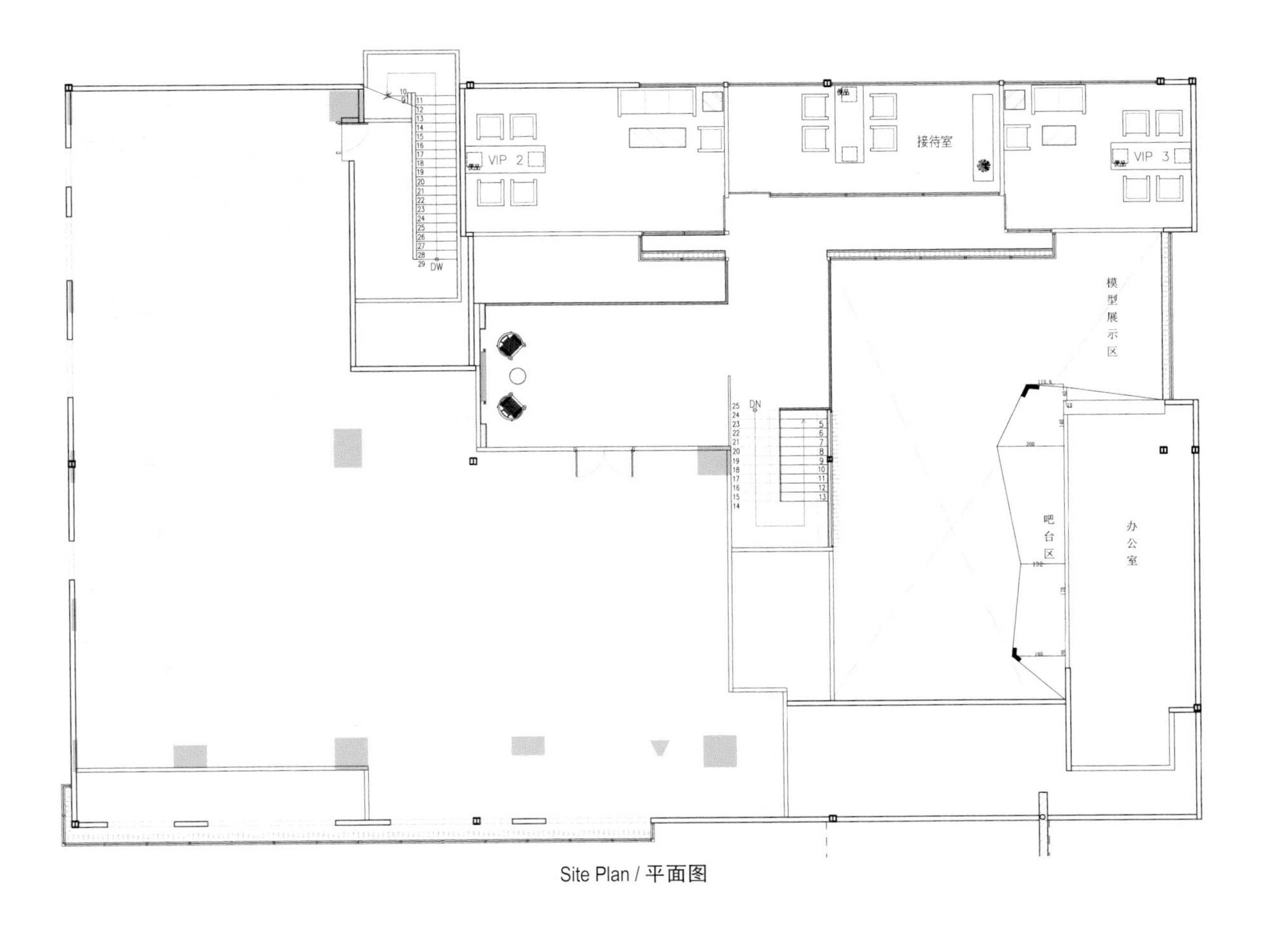

Site Plan / 平面图

THE
PUHUI
The charm is blooming

The charm is blooming

THE
PUHU
The charm is blooming

Putian ECO Sales Center

莆田 ECO 城营销中心

Design Company: North Coast Design
Designer: Wang Jiafei
Photographer: Wu Yongchang
Area: 200 m^2
Main Materials: Imitated Stone Brick, Technology Board, Green Grass

设计公司：福州北岸室内设计有限公司
项目设计：王家飞
项目摄影：吴永长
项目面积：200 m^2
主要材料：仿石材砖、科技板、绿色草皮

The design of ECO city follows the style of the series. Mixing of modernization, sense of city and green elements, with a large space and height, bring the outside view into the interior space. It makes a harmonious combination of decoration of indoor and outdoor. Cream-color is the main tone of the decoration, showing the whole style and impression to visitors.

The sense of architectural body matching with streamline builds a grand modern exhibition space of sales center. Materials and structure are made according with a specific rhyme scheme, to describe the design elements.

The nave of building is spacious, and the assembled point structure in chating room matching with perspective glass wall strengthen the dynamic visual sense of the sales center. The cream-colored and black cloth sofa bring comfortable sense to visitors. Water-drop shaped ceiling lamp creates a cross of points and plane with wide space of

the architecture. The light and massive elevation forms a unique scenery of light and shade.

Without crystal lamp and Rome posts, only flowing line, soft light, dynamic furniture and simple texture present a non-flashy scene. Green grass are used to make the concrete space more livable. It balances the humidity and temperature of the air. Plenty and variety design element reduce the complexity of the environment. The noisy is filtering while the peace is saving.

ECO城营销中心的设计沿袭了楼盘的系列风格，将现代感、城市感与绿色元素穿插其中，利用空间的大体量和高度，使室内空间呈现室外建筑景观般的场景，形成与外景相辅相成的室内外装饰风格。在色调上采用典型装饰主义的米色调，使来访者在营销中心就可以感受到楼盘的整体风格和形象。

ECO城营销中心结合建筑结构，将强烈的体块感与灵动的平面流线结合，营造出恢弘大气的现代售楼展示空间。材料的组合与体块的构造，不是简单的拼贴，而是以某种特定的韵律和节奏，对空间设计元素进行描述。

设计师在空间布局上一气呵成，中堂宽敞开阔，一侧的洽谈区以聚集的点状形式，配合玻璃墙面的透视效果，加强了营销中心的视觉动感。米色、黑色的布艺沙发让人倍感亲切，配以相同色调的仿石材砖，使得空间色彩协调。垂落的水滴吊灯，巧妙地利用建筑空间的空阔感，在点与面的交错中，产生错落有致而又朦胧的视觉效果。在灯光的配合下，洽谈区块状的立面形成独特的光影风景。

没有中空的水晶吊灯，也没有欧式的罗马立柱，有的只是流畅的线条及柔和的灯光相互穿插，动感细腻的家具与简约纹理的科技交织一起，映射出的不再是浮华。绿色草皮的使用让钢筋水泥般的空间顿时有了呼吸感，不仅平衡了空气中的湿度和温度，也具有很强的设计理念与创意。丰富而变化的设计元素弱化了环境的繁杂，仿佛所有的声音都被过滤掉了，剩下的只有城市的宁静和安然。

Victory Garden Sales Center

凯旋荟销售中心

Design Company: Guangzhou Intergrowth Form Engineering Design Co., Ltd.
Designers: Peng Zheng, Xie Zekun
Main Materials: Marble, Granite, White Lacquer Board, Wood Mixed Spell Finishes, Mirror Stainless Steel, Glass, Carpet

设计公司：广州共生形态工程设计有限公司
项目设计：彭征、谢泽坤
主要材料：大理石、麻石、白色烤漆板、木杂拼饰面、镜面不锈钢、玻璃、地毯

The project is located in beautiful scenery of Dinghu mountain scenic area. The designers take the advantage of the terrain conditions, fully tap the internal local spirit, create a unified and winding space with unique style. It's designed to blend the building into the local environment and landscape.

The sales center is decorated with stones which will be harmonized with the mountain. The single space is reorganized by designers. Its folding shape and quality feel are joining with the appearance perfectly. The furniture inside meet the functional requirements of people, and go well with the speciality of space. The outdoor landscape square, as an extension of the sales center, will brings people's view back to the unique geographical environment.

项目位于风景优美的鼎湖山风景区，设计利用了现有的地形条件，充分挖掘当地的内在精神，并受此启发创造出了一个内外统一、曲折有度的独特空间，旨在将当地优美的环境和景致融入到整个建筑空间之中。

营销中心的设计巧妙地应用了山形折面，以"石"对应山，突出了项目优越的地理环境。室内空间将原本单一的空间分解、重组，其折叠形式与质感呼应建筑的外部特征，带来连续的空间体验。室内的家具在满足功能需求的同时，也如雕塑般与空间特质相得益彰。户外景观广场作为营销中心功能的延伸部分，将人们的视线带入项目独特的地理环境中。

凯旋苍
VICTORY GARDEN

洽谈区
Negotiate
Area

卫生间
Toilet
电梯间
Elevator

凯旋荟
VICTORY GARDEN

Zhongqi Green Headquarters Club

中企绿色总部会所

Design Company: Guangzhou Intergrowth Form Engineering Design Co., Ltd.
Designer: Shi Hongwei, Peng Zheng
Area: 1,000 m^2
Main Materials: Marble, Composite Solid Wood Floor, Grey Mirror, Stainless Steel, Light Stone

设计公司：广州共生形态工程设计有限公司
项目设计：史鸿伟、彭征
项目面积：1 000 m^2
主要材料：大理石、复合实木地板、灰色镜面、不锈钢、透光石

The building Zhongqi Green Headquarters is a complexed building assembles office work, exhibition and company club. It has architectural functions of compound and diversification. The word "green" means a concept of "Ecological Office Work".

中企绿色总部，其建筑功能具有复合性和多元化的特征，集办公、展示和企业会所等多种功能于一体。所谓"绿色"则是开发商倡导的一种"生态办公"理念。

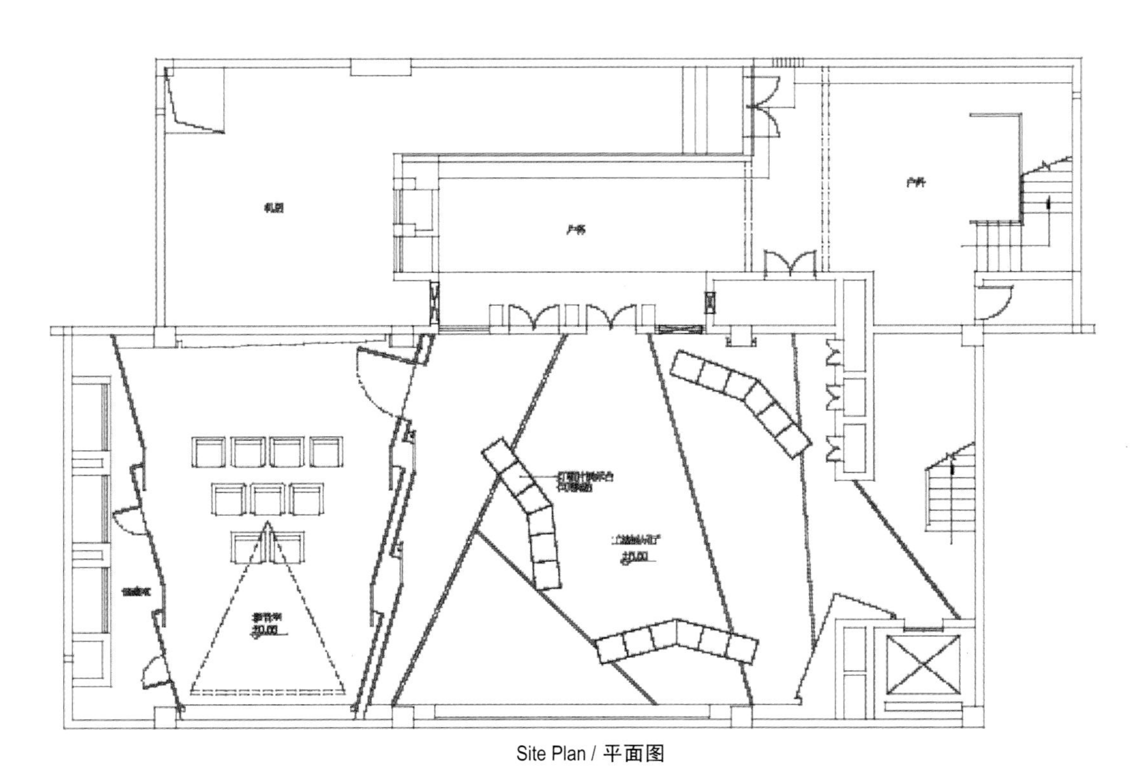

Site Plan / 平面图

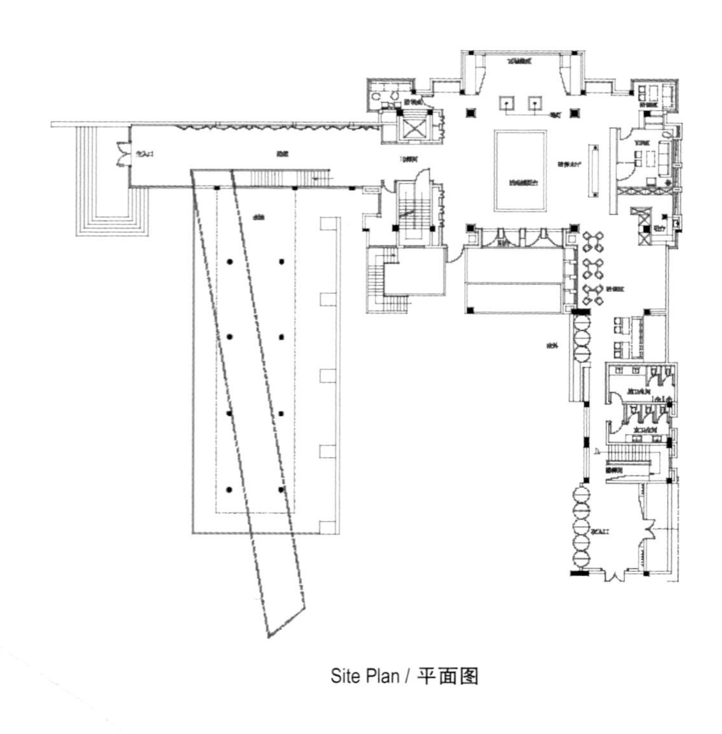

Site Plan / 平面图

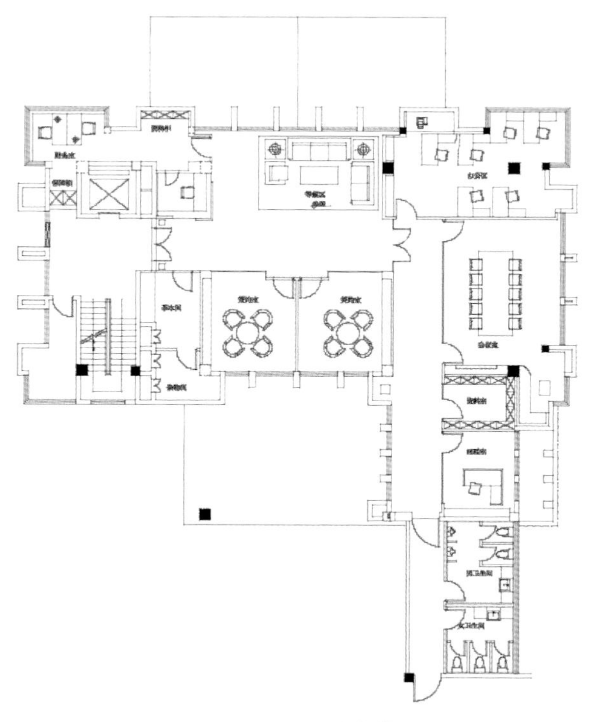

Site Plan / 平面图

LADY

Langrun Garden Sales Center

朗润园营销中心

Design Company: YuQiang & Partners Interior Design

设计公司：于强室内设计师事务所

A spatial setting facing to the courtyard with flickering tree lines is created by introducing such outdoor natural elements as wood grain, green color and so forth into the interior, and by the use of the geometric cut-out patterns and soft lighting. The pleasant surprise brought about by this placid contentment engenders a feeling of authenticity abundance with intimacy to Mother Nature. This has brought along a natural spatial realm of uncluttered simplicity, comfort, and all-season greenery to a northern city.

通过原木、绿色等元素的应用将户外树木的自然氛围延伸至室内，利用几何镂空的图案造型、柔和的灯光，营造出一个面向庭院、树影摇曳的空间。以此带来一份平淡的惊喜，感受一种亲近自然的真实。给北国的城市带来一个简单、舒适、四季常青的自然空间。

Jane Eyre
Jane Eyre
Jane Eyre
Jane Eyre
Jane Eyre

Xingfuli Sales Center

幸福里售楼处

Design Company: Crox International Co., Ltd.
Designer: Tsung - Jen Lin
participants: Li Bentao, Han Qiang, He Shan
Photographer: Wang Jishou
Area: 760 m^2
Main Materials: Recycled Wood, Cement, Black Glass

设计公司：阔合国际有限公司
项目设计：林琮然
参与设计：李本涛、韩强、何山
项目摄影：王基守
项目面积：760 m^2
主要材料：回收木材、水泥、黑玻璃

For the successful implementation of such a green zone for innovations, Capital Domain commissioned avant-garde architect of the natural styles, Tsung-Jen Lin to create the showroom that will symbolise Lahas Zone. Differing from past sale centers, Lahas Zone was comprohensively renovated from an old factory instead of built a new on the Lahas Zone site in accordance to maintaining green. In facing the many challenges of the old structure while maintaining the integrity of recycle design and adhering to the developers intentions, Lin has created Chinese stylised models of living for the post contemporary age. Lin Believes that the principles of green design is in the recyled use of the property and how to utilise an abandoned building undergoing a program change to it's full potential.

The design for this abandoned factory would also have to consider long term use beyond just a showroom, to incorporate a club like setting for future use in the effort the be true to green design. The plan is to continue transforming the space's fucntionality in the future, section by section, to maximism its recyle output and a continuing renew to program. However the current main purpose was to utilise minimal construction to achieve a showroom with an air of "Subtle Happiness". (The idea of "Subtle Happiness" or "Syoukakkou" cames from the famous author Haruki Murakami to mean: Though it's subtle and maybe pedestrian, it's a feeling of happiness that you're completely sure of.) They are the key point of the design that considered by Lin.

To ease the cold atmosphere of the building's bare cement facades, Lin tried to aesthetically create a space thats in contrast to the building but would fit the warmer atmosphere that humans need Therefore by adding wooden elements upon the bare concrete, combining tonality of the materials it casts rough and delicate shadows through the space texturising the spacial visage. The combination of these materials used through out the site blurs the thresholds of exterior and interior while it breaks through the boundries of architecture and interior space. The layout had to be displayed well to not only to house the showroom models for the Lahas Zone but also display the exterioral relationship between the unit models within the sunny atrium. To feel the nature from within the unit models, the atrium had to simulate the desired neighbourhood surroundings inclusive of backyards, veranders and vertical gardens etc. thus to better portray the green life to customers; to experience outside from within and vice versa.

As the feature of the atrium, to welcome guests and customers, erects a giant wooden cavern structure, acting as the focus. The structure starts way before one even enters the space, drawing people into the interior and then attentions will be paid from the floor all the way to the 3 storeys high ceiling that folds over to represent the roof to truly display a crossing of the interior and exterior threshold. The " roof " is refracted in to many slices to reflect the atriums intake of sunlight, encouraging the Chinese people's traditional views on wood and warmth, which is perfect for an atrium where people can wait or discuss the Lahas Zone. The other section of the atrium that is not encompassed by the wooden " roof " but is left with it's original solid concrete walls with the model of this residential developement to give customers a sense of assurence and cofidence. The balance of the atrium between the two distict sections creates a functional and emotional dialogue for customers to decide their dreams for their future lives.

Lin believes that people should persue a more relaxing atmosphere for living, thus monochromatisied the showroom, to depict a more peaceful life style. The whole showroom is set up like a little cafe where people can come and enjoy a drink or some coffee, to read a good book or some magazines under the warm sunshine. The subtle mood of the space is set for future occupants to project their dream lives of living and working into the Lahas Zone, to own "Subtle Happiness" for themselves.

为确保成功执行绿色创意理念，大都置业委托设计风格前卫、自然的建筑师林琮然，打造义乌第一个绿色销售中心。该项目不同以往的销售中心，它没有选择在基地内建造新建筑，而是寻找废弃的厂房，将它进行全面改建。面对原空间结构上的种种限制，建筑师坚持再利用的理念，结合开发商的意向，创立一个全新的中式典范。林琮然认为，绿色的观念在于持续的经营，面对一个闲置的建筑，应思考如何去再设计并重新利用，如何进行功能转换，把空间的使用价值发挥到极致。

旧厂房华丽转身，成为全新的销售中心，并且将会所的功能隐藏其中，设计中更有可持续性的长远规划，所以设计工作本身就意味着环保。设计随着时间的推移能够实现空间功能规划的转换，分期、分阶段的布局，让空间保留最大可能性。尽可能以最少的施工，完成目前阶段的使用需求，并努力创造一种“小确幸”的参观体验（微小而确实的幸福，出自村上春树的随笔，由翻译家林少华直译而进入现代汉语。），这都是建筑师构思的设计重点。

首先，为缓解裸露水泥表面的冰冷气氛，林琮然试图从美学角度打造有着对比度高及人文品位的空间。因此在原始水泥质感外添加了温润的木头，两种不同调性的元素在此结合，让建筑表面产生粗糙与细腻的光影变化，视觉层次丰富；水泥和木材在空间内外相互越界，既模糊又突破了建筑景观与室内空间的界限。平面布局也延伸该理念，为配合内部三套复式样板间的对应关系，在建筑内部设置了充满阳光的大厅与垂直绿化的中庭，将自然延伸到室内，重新创造出虚与实相构筑的环境。进入样板房之前的走廊，展现出如里弄般的气氛，也让销售中心呈现出一个微型社区独有的生活化和人性化的参观体验。

此外，接待大厅为凸显空间张力，植入如自家屋顶的大型木折板，对外巧妙地作为入口的视觉焦点，对内由下而上延伸至天花，并蔓延户外景观，使人可以由庭院慢慢进入建筑，将内外环境真正融为一体。而偌大的大厅也被木折板区分成两个区域。能够反射阳光的多面木板可引发中国人对木头与家庭的情感，该区域作为交流等候区来使用; 另一区域由水泥构成，保留地面与墙面坚实稳重的原始韵味，借由模型沙盘强化了人们勇于追求未来的信念，两者的交汇重新塑造出简约而富有现代感的大气空间，通过二元一体的设计达到功能与情感的完美对话。空间在石与木、虚与实、新与旧间找到更为内敛、微妙的共生，如此富有实验性的生机，再一次完美诠释了前卫、自然的设计理念。

林琮然希望人们能以更加从容的姿态去追求未来，因此销售中心的商业色彩被刻意弱化，设计中更多融入细腻的人性化需求。空间在情境上的转换，有助于引起人们对人生的思考，进而让生活与工作在这里编织出新的梦想。人生要一点一滴抓紧那些微小但确实的满足，“幸福里”就是为那些“小确幸”所设计的。所以这里更像是一个咖啡馆或者小酒馆，期待在这里，可以舒服地坐在沙发上，品尝着拿铁，望着窗外的风景，享受温暖的阳光，信步于青草地，或靠在树下读一本好书，简简单单地想象未来的情景，带着村上春树的风度，且听风吟。

Dalian Vanke Cherry Garden Sales Center

大连万科樱花园销售中心

Design Company: YuQiang & Partners Interior Design

设计公司：于强室内设计师事务所

The conceptualization here evolves from the cherry blossom. The handling of the spatial aspect has succeeded in breaking free from the dwelling's inherent "boxy" appearance by adopting a set of polygonal outlines to traverse and produce a decomposition of the spatial expanse, as the zigzagging turns of the geometric forms and interfaces form an inspirational response to the enchanting sensation of the surrounding environment's embrace of lush green hills. A color palette extends and enhances the elegant white and pastel pink vista of cherry blossom from the window, with a set of snowy-white striated patterns, hardwood outlines, and light grey leather covering to be complemented by matching natural wooden chairs to convey the ecology concept of the design, ensuring the spatial ambience is even more intimately integrated with Mother Nature.

本案以樱花为元素，展开构思。空间上，打破原建筑固有的“盒子”形体，采用折线穿插以分解空间。抽象的几何形体，界面的转折起伏，与环境中重山环绕的灵动感相呼应。色彩延续了窗外樱花高雅的白色与粉色，细纹雪花白、实木线条、浅灰色皮革配以原木座椅，彰显的设计生态理念，使整个空间氛围更加贴近自然。

Tianxia International Center Sales Center

田厦国际中心销售中心

Design Company: YuQiang & Partners Interior Design

设计公司：于强室内设计师事务所

A suspended design form that makes use of geometric combination of hexagonal shapes is extending to full of the whole space, and by superimposing it with a mass of fine lines resonating the application of further linear elements in the black-and-white wood grain texture on the stone floor tiles has allowed this interior space to retain the inherent rational quality of the building whilst also imparting it with a certain subtle ambience. The series of geometric cut-out patterns reaching down from the ceiling serve the purpose of subdividing the space into two distinct principle zones, and the end result is an open area that also gives thoughtful considerations to the user's privacy. The embellishment with metallic fixture and fitting is bringing a particularly trendsetting appeal to this space.

利用六边形组合成的带有几何感的悬吊造型充满了整个空间，密密的细线叠加在造型上，黑白木纹理石的直线元素运用于地面，使得室内空间既保留了建筑自身的理性又增添了些许柔美的气氛。几何镂空的图案造型作为分隔空间的装置，将洽谈区与展示区分开，既开敞又不失私密感。金属材质的点缀，则更增加了空间的时尚气息。

Tender Luxury

温柔的奢华

Design Company: MoHen Design International
Designer: Hank M. Chao
Participants: Wang Yingjian, Hu Xinyue
Photographer: Maoder Chou
Area: 700 m^2
Main Materials: Oak, Beige Trarertine, Glass, Iron Pieces, Poplar

设计公司：牧桓建筑 + 灯光设计顾问
项目设计：赵牧桓
参与设计：王颖建、胡昕岳
项目摄影：周宇贤
项目面积：700 m^2
主要材料：橡木、黄洞石、玻璃、铁件、白杨木

The project is located by the side of the Huangpu River, Pudong, Shanghai, is a new developed land surrounded by well-planned landscape. In the entrance, the designer deliberately emphasized the depth of field and separated by the pool to cut the moving lines of the rear leading to the toilets. To project the flow texture of the water, designer placed the dynamic projection of water ripples on the toilet wall facade, which have a visual linkage of Huangpu River on both indoor and outdoor view. In the seating area, the designer used bookcase hanging from the ceiling but not to the down floor as a separator, it does not completely block the space, but yet still can have the visual part of the penetrating sense. Metal curtain is also used in this space as a kind of soft segment. Besides, for the bar design by using the practices of the carved handling, with straight lines around which makes a strong contrast. The background served as a foil to the poplar to let the light and shadows looming, inadvertently spilled on the ground to increase the poetic sense, just like a stone in the forest. Walkway leading to the VIP room, designer set the same projection wave on the ground, this kind of dynamic way introduced a more interesting process of entering the room, and also make up the design defect of limitation to "hardware" which can not be interacted with people. The fireplace is also placed in the VIP room to increase the visual sense of warmth.

The overall color presents calm and restrained tone to respond to the concept that deliberately avoided the extravagant sense, but through the material is an appearance of

a more deep gorgeous. The wide amount of high ceiling and the released space is a performance of another gorgeous — the luxury of the scale, this kind of luxury which can not be achieved the visual tension by a small space, as the designer, is trying to convey a visual vocabulary through this project.

本案坐落于上海浦东黄埔江边，是一个新发展的地块，周围有良好的配套景观。入口处设计师刻意强调景深并用水池隔开后方通往洗手间的动线。在通往洗手间的立面墙上以水波纹动态投影展现流水般的质感，让室内与室外的黄浦江有了视觉上的联动关系。座位区则利用从天花垂落下来的书柜作为区隔，但却又不完全阻隔空间，使其有了视觉上的穿透感。金属帘也作为软性区隔方式被应用于空间。另外，以雕刻的手法处理的吧台，与周围的直线条形成对比，背景衬托白杨木的树林里，光和树影若隐若现，不经意地洒在地面，宛若林中的一块石头极具诗意。通往贵宾包厢的走道也同样将水波投影于地面，这种动态的方式让进入包厢的过程有了趣味，也弥补了设计局限于“硬体”，无法与人互动的缺陷。包厢里壁炉的设置增加了视觉上的温馨感。

空间整体色调沉稳内敛，呼应刻意回避的铺张感，但透过材质可传达一种更深沉的华丽。挑高释放出来的较为宽阔的空间量体表现的则是另一种华丽——尺度上的奢华，这种奢华不是一般小空间能够达到的视觉张力，这也是设计师想要透过本案传达的一个视觉语汇。

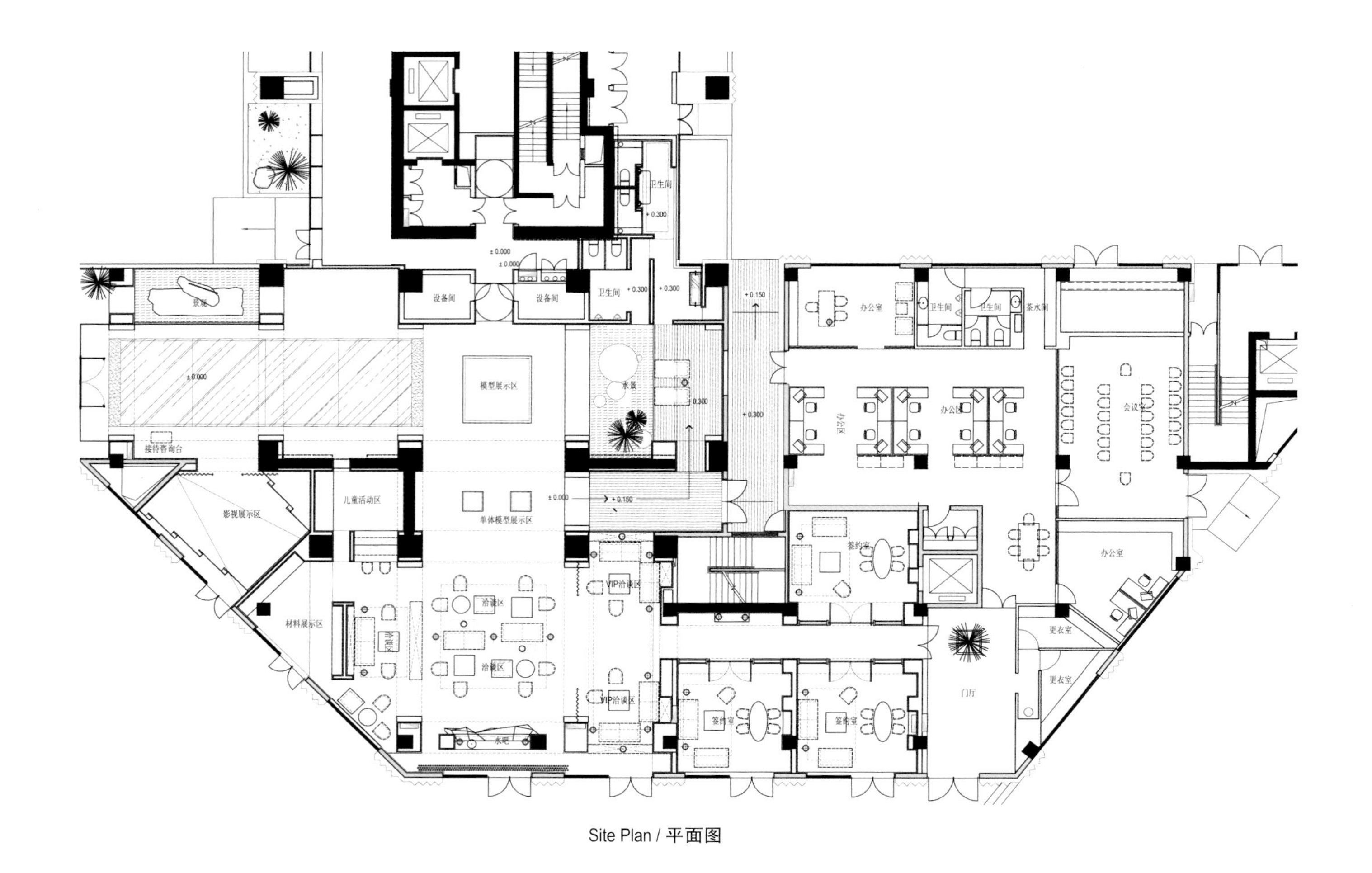

Site Plan / 平面图

Star and Cloud Sales Reception Center

星汇云锦销售接待中心

Design Company: APEX Design and Consultant Company
Designers: Li Qien, Chen Zijun

设计公司：尚策室内设计顾问有限公司
项目设计：李奇恩、陈子俊

The project is located in Qiaokou, district Hankou, Wuhan. It likes a pearl on the side of river, shining with dazzling brilliance. According to special property of this area, designers put the follow concept into the design.

Water is the theme of this design. A water stylished crystal lamp and elliptical background wall matched with wave modelling wall. The integration of light and scenery makes space changing. Visual decorations well combined with elements of "star and cloud", constructing the internal space of meditation.

The building is divided into two layers. In the first floor, there are two luxury VIP guests rooms, 7 meters high audio room, six independent VIP lavatories, and fashion children's play room. These are presenting a noble quality to visitors. In the second floor, there are office area and contract signing area. The unique design presents a harmonic atmosphere, to exalt the theme of harmony.

As a sales center, it presents the internal quality, noble, fashion of Qiaokou golden triangle project, and show a image with strong sense.

星汇云锦销售接待中心位于武汉市商贸中心地区——汉口硚口区。硚口这颗汉江边上的明珠，正闪耀着夺目的光辉。根据建筑所处的特殊场域，设计师引入了以下设计理念。

设计以水为灵感，销售大厅洽谈区上空由水的曲线形态演变而来的水晶灯与椭圆的金钢背景墙及两端的波浪造型墙相呼应，光的透入与景的融合使空间意义发生了变化，不仅仅只是富有视觉力的饰物，同时也很好地结合了楼盘“星汇云锦”的元素，构筑出这个引人冥想的内部空间。

该中心分为两层，一层设有两间豪华 VIP 贵宾室、挑空七米高的椭圆形影音室以及六个独立的 VIP 洗手间，同时还有充满活力的儿童空间，这些足以给来访者带来高贵尊崇感。二层为认购签约区及办公区域，与众不同的设计，很好地调和了现场气氛，表现出和谐尊崇的主题。

作为销售中心，本案从某种意义上来说代表着硚口金三角项目的内在气息，尊贵、时尚，并且有着强烈的形象感。

Heidel Berg Sales Center

海德堡售楼处

Designer: Wang Wuping
Area: 450 m^2
Main Materials: Teak, Marble, Leather, Grey Mirror, Wallpaper, Artistic Coating

项目设计：王五平
项目面积：450 m^2
主要材料：柚木、大理石、皮革、灰镜、墙纸、艺术涂料

The targeted consumers of this project are city elites. It has both commercial and residential functions. It requires small space to present a high level quality. So, the design of sales center is very important.

Grand visual effect of space and strong contrast of colors build a tasteful and delicate atmosphere to deliver a powerful, trustful and good quality sense.

There are reception center, sales and control center, chating area, depth discussion area, display area, VIP room, multi-media room, and room for marketing director in this sales center. In the hall, the central decorated with the waterscape, wooden line decorated wall with the letter "H", "D", "B" on it deliver the cultural concept of company and strengthen visual impression.

本案服务于城市精英阶层，有商业，有住宅，虽然占地面积不大，但力求打造成城市的高端精品项目。所以，销售中心的设计也就显得尤为重要。

整个销售中心呈现出现代、大气的空间视觉效果，强烈的色彩对比，营造出一种精致的销售氛围，给客户传递一种实力、品质、信任的感觉。

在空间规划上，本销售中心设有接待处、销控中心、洽谈区、深度洽谈区、展示区、VIP 室、多媒体室等功能空间，大堂正中央设有水景，两侧墙体以木线条元素为主，同时点缀“H”“D”“B”等英文字母，既传达了公司的文化理念，又增强了视觉感染力。

DB

HDB

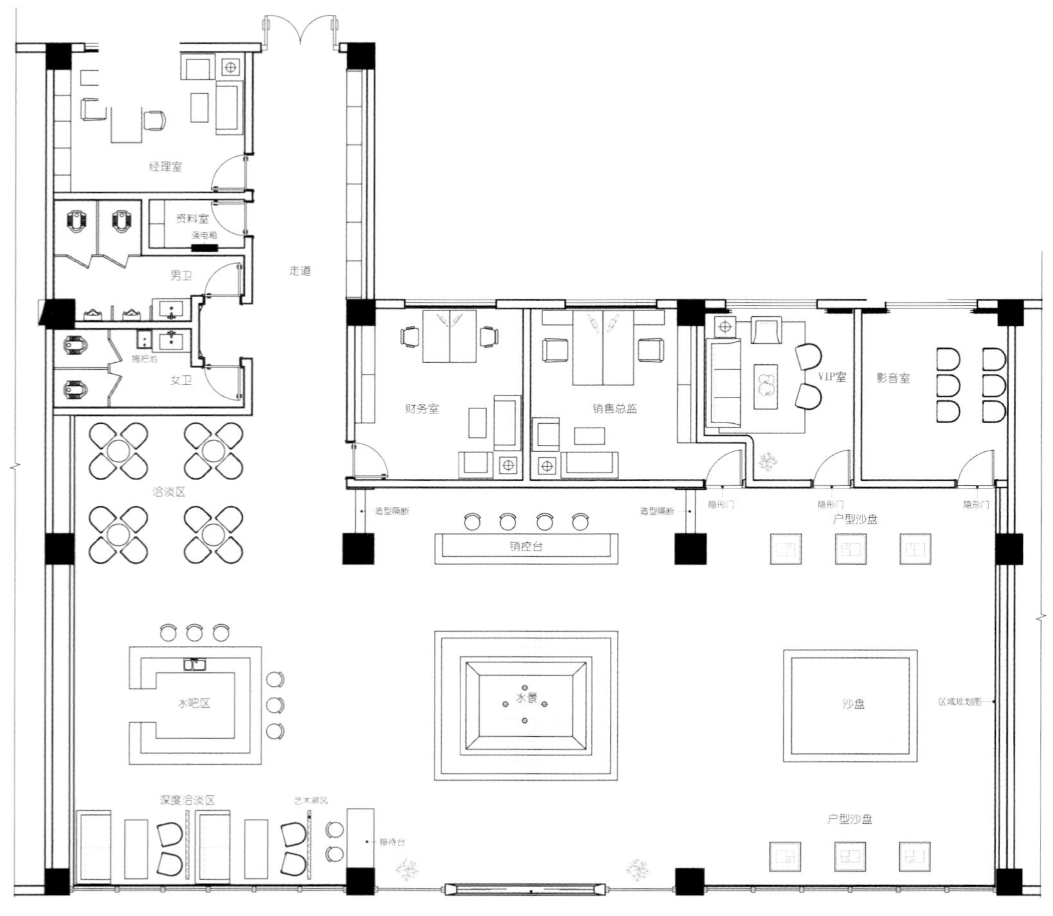

Site Plan / 平面图

COFCO Jinyun Sales Center

中粮锦云营销中心

Design Company: Hover House
Designers: Nie Jianping, Zhang Jian
Area: 1,000 m^2

设计公司：深圳市世纪雅典居装饰设计工程有限公司
项目设计：聂剑平、张建
项目面积：1 000 m^2

The sales center has responsibility of presenting the quality of the project and brand identity. It shows the territorial advantages and the delicate brand identity of COFCO. The design focuses on the expression of the smooth and grand quality of space, and introduces fashion elements to present sense of contemparaneity.

The concise design of section and the vertical lines make the space more rhythmed. The floor with liana pattern, the large wall painting and jumpy musical notes, containing the connotation of COFCO identity.

Crystal lamp, ornamental bar, crystal glass sticks are all designed to show unique ingenuity, adding unparalleled decorative effect and artistic taste.

营销中心担负着展示项目品质及品牌形象的职责，要体现楼盘所在地区的优势，同时也要反映中粮集团精致的品牌形象。设计在强调整体空间流畅、大气的同时，时尚元素的融入表达了其与时俱进的时代感。

简洁、利索的切面与贯穿始终的竖线条盈于整个空间，使空间充满节奏与韵律。植物藤蔓的地面拼花与巨幅的墙面艺术壁画，恰似乐章间跳动的音符，蕴含着“中粮”的品牌内涵。

水晶吊灯、吧台镂花、水晶玻璃棒吊饰都是经过独特设计的艺术品，为整个空间增添了独一无二的装饰效果，也提升了空间的艺术品位。

中粮
COFCO

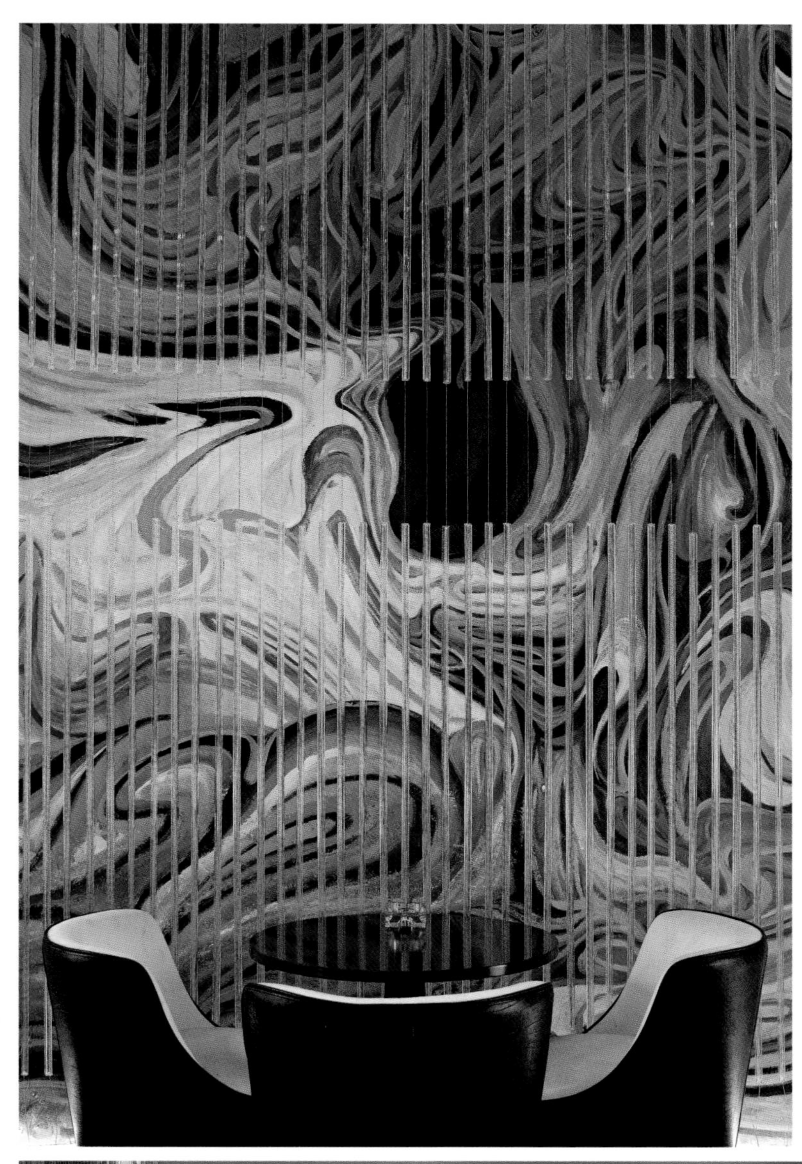

西乡人民医院
宝安体育馆
宝安中心站
宝安区政府
图书馆(在建)
大铲湾
珠江口
前海湾

Site Plan / 平面图

Guigu Reception Center

贵谷售楼部

Design Company: LKK' Design Co., Ltd.
Designer: Lin Kaixin
Participant: Hu Chenyuan
Photographer: Wu Yongchang
Area: 400 m^2
Main Materials: Antique Brick, Teak, Texture Coating

设计公司：福州林开新室内设计有限公司
项目设计：林开新
参与设计：胡晨媛
项目摄影：吴永长
项目面积：400 m^2
主要材料：仿古砖、柚木、机理涂料

ikk design
ikk design
ikk design

贵宾室

Jinti Sales Reception Center

津提售楼处接待中心

Design Company: Danny Cheng Interiors Limited
Designer: Danny Cheng

设计公司：郑炳坤室内设计有限公司
项目设计：郑炳坤

Mountain and City Sales Office, Chongqing

重庆山与城销售中心

Design Company: One Plus Partnership Limited
Designers: Law Ling Kit, Virginia Lung
Photographers: Law Ling Kit, Virginia Lung
Area: 1,600 m^2
Main Materials: Roller Blind, Carpet, Mirror, Glass, Marble, Corian Stone, Paint, Plastic Laminate, Outdoor Timber Decking, Stainless Steel, Wallpaper

设计公司：壹正企划有限公司
项目设计：罗灵杰、龙慧祺
项目摄影：罗灵杰、龙慧祺
项目面积：1 600 m^2
主要材料：卷帘、地毯、镜、玻璃、大理石、可丽耐石材、油漆、防火胶板、户外木、不锈钢、墙纸

Gazing across the river from urban Chongqing, here come the famous scenery district of Nanshan. The green is weaving into the mountains here, with the meandering river, they are all embracing the hilly Chongqing City with their beauty. The sales office of Mountain and City is just located at the picturesque Nanshan district, which the view has inspired the designers for their idea of the interior.

Just as the geographical character of Nanshan, the landscape inside the sales office is assembled with mountains and valleys as well. The walls are conglomeration of triangular planes and oblique line sets, in different tones of gray, the outcome is a powerful and dynamic topographic map. No matter where they place themselves, visitors could feel the embrace by hills exactly like the city itself. While standing on the marble floor that plays a role as the valleys in Nanshan, the energetic ambience of the mountains would stun into visitors retians in the meantime. Whereas, the same geometric of the walls has extended to the grand marble floor, the irregular triangular patterns with various kinds of marble have been well arranged in the same manner. Together, both the vertical and horizontal planes have visualized the mountainous beauty scenes of Chongqing.

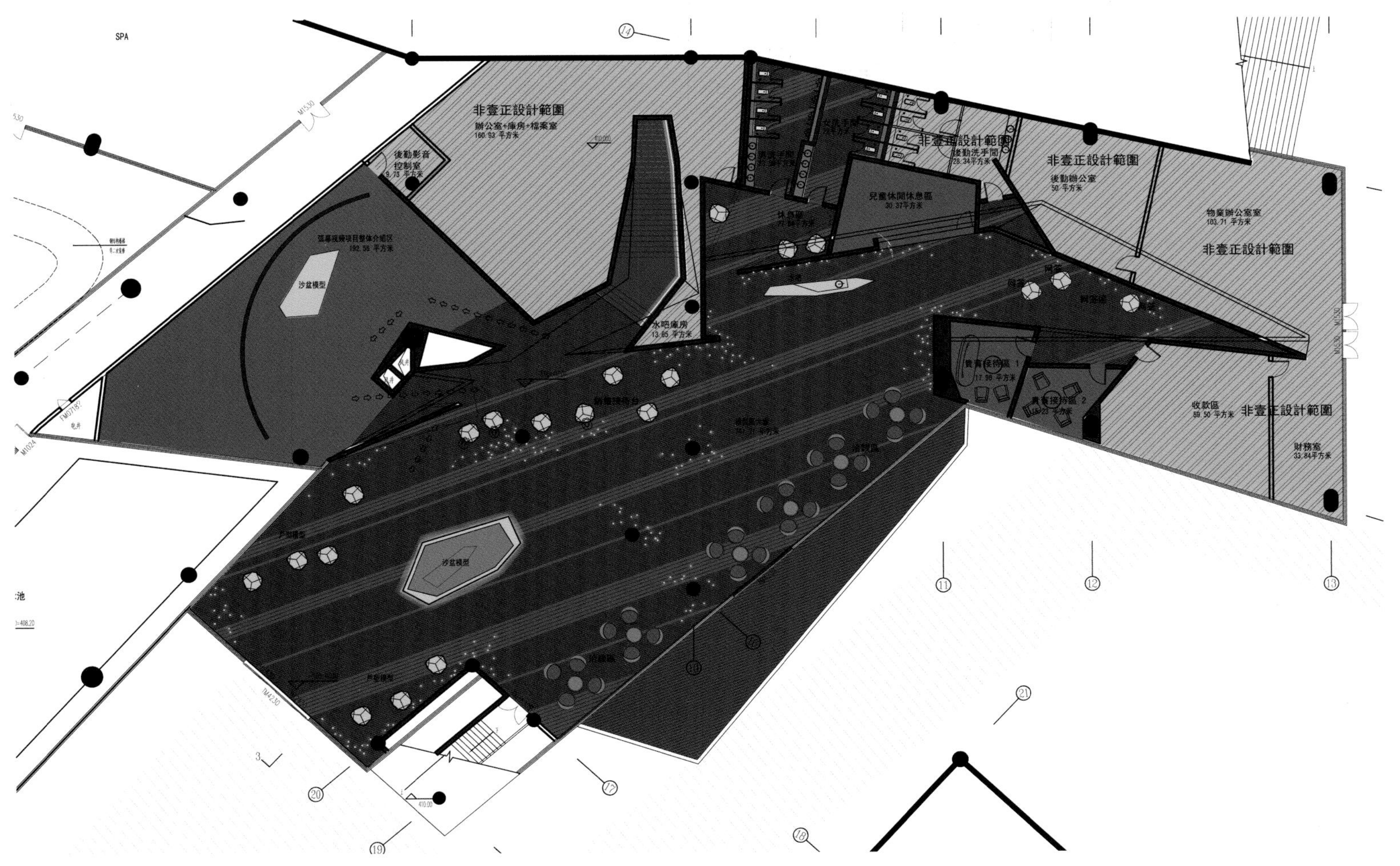

Moreover, even the array of brown stainless steel counters each has an individual form; yet they are in placed in faultless range. They can be seen as a series of monstrous rocks, as well as they can be described as the crystallized form of the cultural spirit of Nanshan. Meanwhile, the eye-catching counters act as the focus points in the grayscale surroundings.

Walking toward the midst of the mountains, a cave appears beneath the cliff ahead, which is the passage to the other floor level. In order not to arouse puzzle feeling among visitors, thus there is a long strip of light laminating along the tunnel. The light transfers the geometric style to the other corner of the interior while easing the depression in the dark.

The numerous strings of LED chandeliers to offer the dreamy mood of rainy Great Southeast. The softness that presents by the rain also forms a contrast with the strong and bulky atmosphere of the "hilly" theme of the entire interior, which is also kind of buffer to visitors'eyes. Take a glance upon the "sky", not only rain, the space also presents visitors a trip to the galaxy; through the celestial shower that shines above with programmed pattern, people's mind could be relaxed and rinsed as well. Within the sky and the ground, among the mountains and the valleys, every little thing within are portraying the spiritual and robust sight of mountainous city.

从重庆市中心向南隔江相望，是有名的南山风景区。此处群峦叠翠，与大江一起环绕山城。重庆山与城销售中心恰好位于山河庇荫的南山区，秀丽景色理所当然成了设计的灵感来源。

一如南山地形，销售中心的室内空间亦由山岳幽谷构成。各处墙壁由深浅灰色的三角及倾斜的线条组成，建造出一幅充满力量及动感的山势地形图。无论室内还是室外，访客皆被众山环抱。然而，站在山谷的地坪之上，脚下也能感受到南山数十山峰的活力。石材地坪饰面也沿用了墙壁间隔的图案式样，以不同角度排列的石材组成大量不规则的三角形。两者共同展现了山城大地活跃而秀丽的景象。

此外，排列于大堂中央的棕色不锈钢柜台，形态各异，整齐有序。可将它们视为嶙峋怪石，也可当作南山大地人文精神之结晶。同时，醒目的柜台也在以灰色为主的室内环境中产生了点睛之效。

游人一路走到“山中”，到了“峭壁”之下，便可直达“山洞”，这也是通往另一楼层的通道。为了让游人不被曲折的山洞石壁吸引、迷惑，长长的条状灯光贯通整条通道。使人消除阴暗中的沉闷之余，也将几何风格引至室内其他角落。

西南地区雨丝婆娑的诗意风景通过一串串 LED 吊灯表现出来，灯雨带来的轻软柔和之感，不但缓和了室内群山的刚强坚实，也为人们的视线作了缓冲。再往天上看，随程序闪烁的吊灯一如星雨下凡，一时间游人恰似随串串星雨漫游于星汉之间，心里也仿佛被洗涤般放松。天与地、山与谷，最终描绘出山城那充满灵气和生命力的秀美景色。

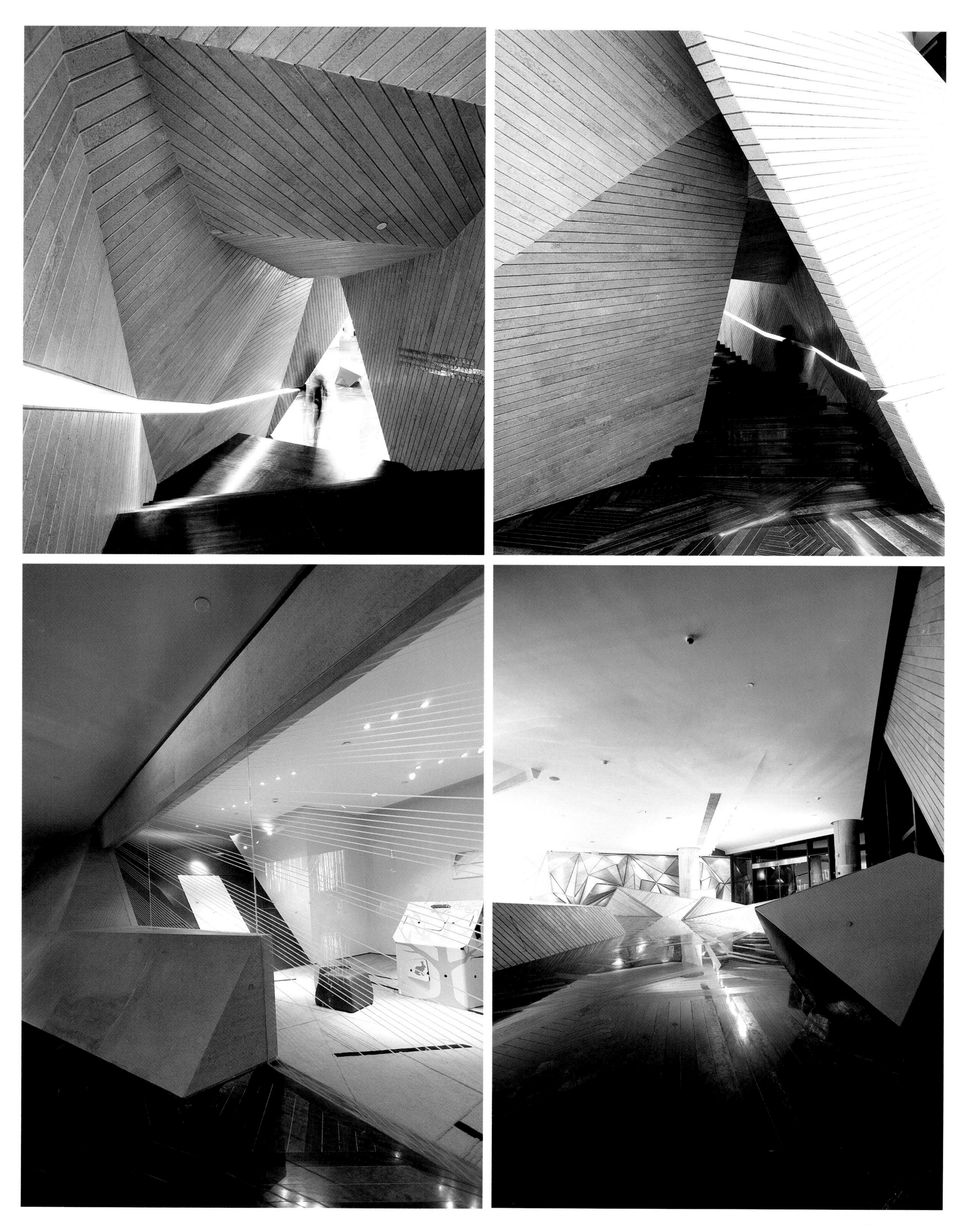

Zhonghang Feicui City Sales Office, Xinjiang

新疆中航翡翠城中心会所

Design Company: PINKI Interior Design I Liu & Associates Interior Design Co., Ltd.
Designers: Liu Weijun, Liang Yi, Yuan Chaogui
Area: 2,300 m^2
Main Materials: Art Paint, Oak Panel, Teak Flooring, Marble

设计公司：PINKI 品伊创意集团 & 美国 IARI 刘卫军设计师事务所
项目设计：刘卫军、梁义 、袁朝贵
项目面积：2 300m^2
主要材料：艺术涂料 、橡木面板、柚木地板 、石材

This project is a high-end club for sales, leisure, entertainment and fitness. The interior design style of this project is American flavor, which highly keeps in coordination with that of the architecture. In the respect of the space plan, according to the space characteristics of the building itself, this case make the reception hall as the center after further modification , thus widely forming two axes which is vertical and horizontal to the American gateway arch hole in a unified form in a large scale, linking each functional areas in the club. The central axis reasonably and orderly connects all functional areas to form a streamline which is really full of rhythm.

本案是一个集销售、休闲、娱乐、健身于一体的高档会所。本案室内设计的美式风情与建筑风格保持高度统一。在空间的规划上，根据建筑本身的空间特点，加以改造，以接待大厅为中心，形式统一的大尺度美式拱门贯连会所各功能区域，形成纵横向两条视野。中部轴线将各功能区合理有序的贯穿连结，形成一个富有节奏感的流线。

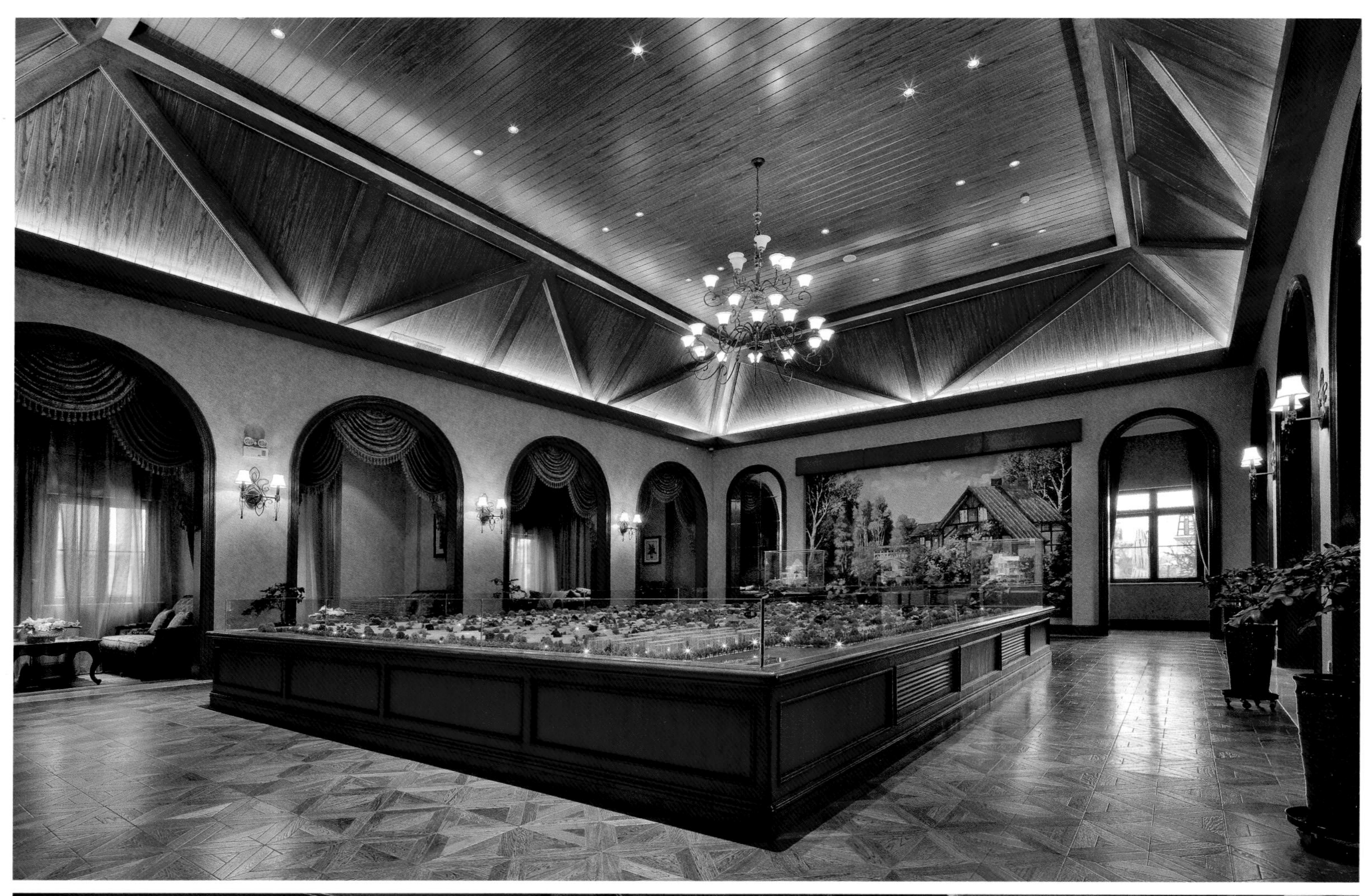

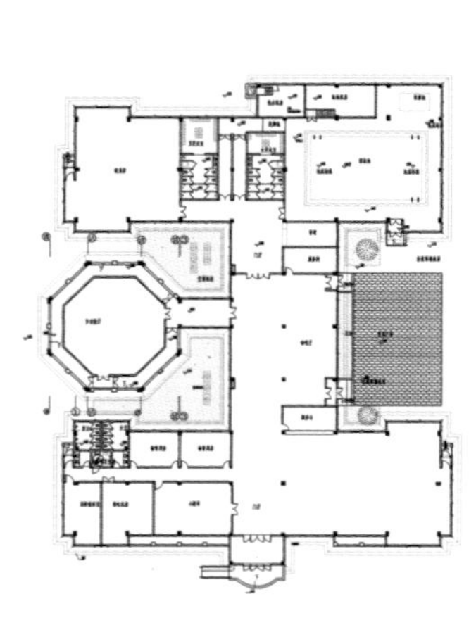

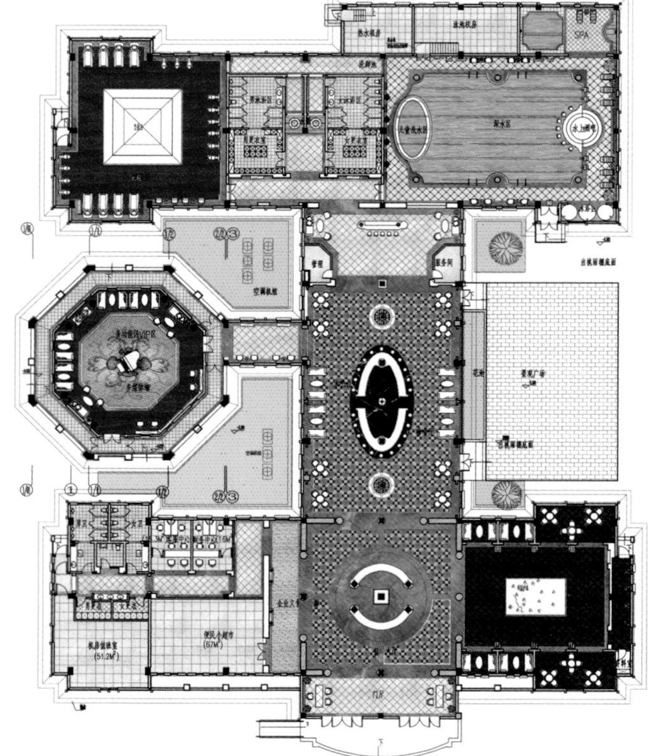

First Floor Plan / 一层平面图

WATER
EARTH
NORTH
AMERICA OR MEXI
MAR
DEL
NORT
MAR
DEL
ZUR
THE PACIFICKE SEA
MAGALLANICA
THE
WESTERNE
OCEAN
THE INDIAN
SEA
THE SOUTHERNE
UNKNOWNE LAND

专注成就专业 品质铸造品牌

华中科技大学出版社建筑分社一直专注于土建类图书的策划、出版、发行工作，出版方向覆盖建筑设计、室内设计、园林景观、城市规划、建筑施工、建筑考试、建筑文化、高校教材教辅等诸多细分领域，部分图书远销海外多个国家和地区，整体实力已跻身全国前列，在读者中享有良好口碑。

经过多年发展，华中科技大学出版社建筑分社已建立起一支精通专业、勇于拼搏、富有激情的团队，并依靠诚信的为人品格、科学的管理方式、先进的经营理念、严谨的办事作风如磁石般吸引了国内外大量优秀作者和经销商，使其图书品质和营销网络在业内一直拥有强大的市场竞争力。

十年树木，百年树人。华中科技大学出版社建筑分社将以传播先进技术。弘扬优秀文化为己任，永不懈怠地为我国出版事业的发展而奋斗。欢迎更多合作伙伴与我们携手共进！

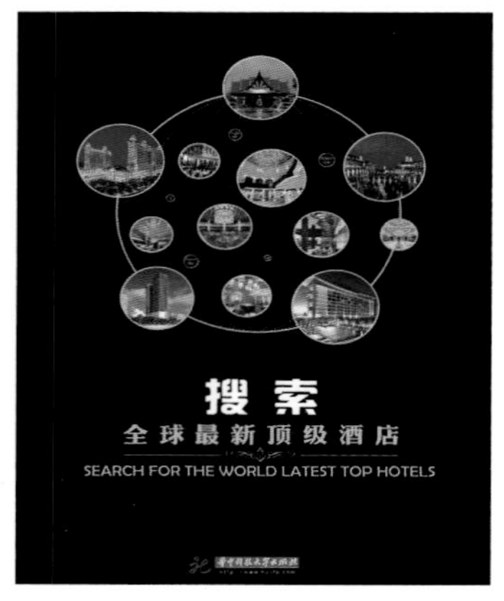

搜索 全球最新顶级酒店
定价：980.00 元

别墅庭院设计 II
定价：358.00 元

越夜越美丽 II 全球顶级夜店设计
定价：338.00 元

极致奢华 居住空间
定价：338.00 元

老有所居 老年人公寓设计
定价：338.00 元

创意办公空间
定价：338.00 元

投 稿 热 线：020-36218949
售 后 服 务：QQ 群号 18943812 / 171468425
总 社 网 站：http://www.hustp.com
分社中文网站：http://www.hustpas.com
分社英文网站：http://www.hustpas.com/en/
电子图书下载：登陆分社中文网站→查看下载流程说明→点击“电子书下载”→进入下载页面→ 下载本书电子版
盗版举报热线：12390 或 010-65212787/2870（全国扫黄打非办）
027-68892461/2429（湖北省新闻出版局扫黄打非办）

网络营销支持：新浪微博 当当网 dangdang.com 亚马逊 amazon.cn 360buy.com 京东商城 淘宝网旗舰店 http://hustp.taobao.com